Putting Soul into Science

Putting Soul into Science

Michael Friedjung

iUniverse, Inc.
New York Lincoln Shanghai

Putting Soul into Science

iUniverse, Inc.

For information address:
iUniverse, Inc.
2021 Pine Lake Road, Suite 100
Lincoln, NE 68512
www.iuniverse.com

ISBN: 0-595-27960-0

Printed in the United States of America

Contents

Preface

This book is the result of thinking for many years about the apparent contradictions between what is accepted as scientific knowledge and various forms of spiritual teaching. These include the "anthroposophy" of Rudolf Steiner, to which I shall refer fairly frequently in the book. These contradictions have preoccupied me since late childhood. Some scientific colleagues are adherents of spiritual movements; they often appear to me to have split their activities and perhaps their personalities in two, without much connection existing between the two halves. The problem is that modern science is based on materialistic assumptions; it supposes that the whole Universe is, in the last instance, explainable by the laws of physics. Such laws are "blind", not requiring any conscious participation. Human beings are then considered to be only very complicated machines. On the other hand, many kinds of spiritual teachings exist. Most religions speak about God or about gods who rule over the world. The existence of non-material components of the Universe like souls, spirits and non-material places that humans can inhabit after death is emphasized in spiritual conceptions, and spiritual development through meditation is sometimes recommended.

Usually only the results of materialistic science based on apparently rigorous methods are considered in the present-day world to be "true", although they seem in many ways to be inhuman, while spiritual ideas, often taught in a dogmatic way, are thought to be mere superstition. Indeed one can argue that differences in religion and religious dogma have in many situations been good excuses for violence and massacres.

It should be emphasized that among spiritual teachers Rudolf Steiner in particular spoke of paths of human development leading to the ability to obtain rigorous scientific knowledge of spiritual truths including, for example, those concerning the spiritual evolution of the Universe and of human beings. This characteristic is certainly a good reason for paying special attention to him. In spite of this side to his teachings, it is still quite difficult to see the connection between what he taught and official science.

The question then arises: is another sort of science possible which does not eliminate the soul and spirit? In this book I look at various basic assumptions of the scientific methods as presently practiced and how they might be changed so as to make another kind of science possible. It now appears clear to me that especially ideas about consciousness, the existence of separate conscious beings and the soul need to be integrated into science; it is not sufficient, as is sometimes done, to make some sort of unorthodox physics if it remains "soulless". In fact, within what is present-day official science it is possible to see that certain great discoveries of the twentieth century can be understood in a different way than as is usual among most scientists. That way, involving the presence of conscious beings with certain soul qualities, is described in this book, as well as why present day physics has something "inhuman" in it. In this work I have been inspired by some of Rudolf Steiner's basic ideas; however, I propose a path for the reader which does not require prior acceptance of the statements made by him or by any other spiritual teacher.

Problems involving human society are also discussed, as they are not completely unconnected with science. I was to some extent also inspired in my approach by activities in political movements of the extreme left in the years following 1968. These activities led me to eventually clearly see the importance of social relations, not only between human beings, but also those involving other types of conscious beings.

I have tried to be as non-technical as possible, only assuming that readers have a certain basic culture and know a little elementary science. The book should therefore be readable by scientists and non-scientists alike. There are very few mathematical expressions. Similarly, the reader does not need to know very much in advance about the teachings of spiritual movements. However, I do expect that he or she be able to think through the various arguments.

Professionally I am an astrophysicist working at the French National Center of Scientific Research (CNRS). In my normal research I study certain types of peculiar stars, such as novae (which have violent explosions), as well as what are called symbiotic stars (each of which contains two stars interacting strongly with each other) and other stars which seem to be surrounded by disks. These objects exhibit many fascinating phenomena and properties not always easy to understand. It is for this reason that they have excited my interest.

I must here give acknowledgement to my parents, from whom I heard about Rudolf Steiner's teachings when I was a child, and especially to my father, Walter Friedjung. Though largely self-taught, he was able to study the spiritual aspects behind mathematics. Part of the manuscript of a book he wrote was published in German in 1968 (Vom Symbolgehalt der Zahl, Europa Verlag, Vienna, Austria). It was through my father that I became interested in science and concerned about how to relate science to spiritual teachings. When I was a teenager I looked for phenomena in astronomy which then appeared to me to be difficult to understand in the framework of official conceptions. Following the influence of my father, I received a scientific education and eventually started doing research in astrophysics. In addition I must mention that my father correlated numbers and human beings in the latter part of his manuscript. This served as an inspiration and starting point for what I have written in this book about soul qualities and consciousness in the eternal world of pure ideas of mathematics.

I wish to thank Daniel Bariaux, who read the manuscript and made many suggestions, Ilunga Mwana Umbela and Irena Semeniuk who, among other things, checked the manuscript for English and typing errors and F.T. Smith for his editorial work. Souad Lebbaz drew the figures. Also I thank E von Bezold, S. Nordwall, G. Zoeller and J. Zorec for helping to elucidate various points.

1

The Role of Present-Day Science—The Aims of this Book

1. *The Importance and Nature of Science*

Modern science plays a major if not a dominant role in the world at the end of the twentieth century. Applications of the results of scientific research have led to the technology which surrounds everyone in developed countries. This technology has very much influenced the daily lives of people. Taking some examples, the energy we use is very often supplied in the form of electricity, this being a product of nineteenth century physics. The electronics derived from more recent physics is necessary for making computers, which not only perform extremely complicated scientific calculations, but also control the operations of many types of machines. Electronics is needed for producing the many kinds of images watched by people, such as those of television and videos. Radio waves, the existence of which was predicted by physics in the nineteenth century, are widely used in communications. Chemical research led to the invention of new substances such as the plastics that surround us in our daily lives. Certain new substances are much used in contemporary medicine. A knowledge of fluid mechanics is needed in the design of aircraft. As a result of such technological developments, it was possible to send men to the Moon in 1969.

The economy of the world at the present time is very much a result of modern technology; the objects produced by the latter and the raw materials needed to produce them play a major role in world trade. It is moreover possible using modern technology to introduce increasing amounts of automation into production, thereby leading to a reduced need for manpower, resulting in unemployment.

Rapid communications are essential for speculative financial operations, which can completely destabilize the economy. Indeed technology does not only produce material comfort and means of communicating and of being informed,

1

but it can also produce new forms of human suffering. The same conclusion is reached even more directly when we think about the role of science and technology in the development of modern weapons.

It is for such reasons that the governments of many countries attach great importance to scientific research. And this research very often needs extremely expensive equipment. In this way we can say that the economy and political decisions are not only dependent on scientific research, but that scientific research is also dependent on the economy and political decisions.

What must be emphasized, however, is that the conceptions and technological successes of science have a strong effect on the way people think. Non-scientists, for whom technology often appears to be a kind of magic, can be led to believe that scientific knowledge is the only reliable form of knowledge. They are told that the scientific method as now applied is, unlike other approaches, rational and rigorous, involving as it does the systematic experimental testing of theories and hypotheses. The application of such well tested theories explains the technical successes. Even though scientific theories change, the basic results of modern science are thought to be true, and therefore to be believed.

The assumptions of modern science are materialistic. This means that if you look deeply enough into the phenomena of the world, they can all eventually be explained by the laws of physics. These laws, as now understood, are not the mechanical laws of nineteenth century physics, so materialism is no longer mechanical. They are very abstract and mathematical, but they are usually understood to be "blind", eliminating the action of any conscious being. Everything cannot be predicted by these laws which, as we shall see, contain the unpredictable; what is unpredictable, however, is thought to be only the result of blind chance. This is extrapolated from physics to other sciences such as biology. In biology, Darwinian natural selection is also blind: species of living organisms evolving by chance are better able to survive in their environment without the intervention of any plan or idea. Competing species less able to survive are eliminated.

It is in such ways that the properties of matter and the world we live in are explained as being the result of very small-scale processes at the atomic and subatomic levels. The structure and distribution of matter in the universe is explained by processes which followed what is called the "big bang", when the universe is thought to have been very small, very dense and very hot, while after the big bang it is thought to have been in continuing expansion. Man, like all other living organisms, is considered to be a kind of machine and his consciousness and thinking are due to the functioning of the matter and the nerve cells of which his brain is composed.

It is in such a context that one may wonder whether spiritual conceptions of any kind—such as the various religions—involving ideas outside physics, are really relevant in the present day world. Or are they only remnants to be swept away in the course of time? Is there a way of reconciling the spiritual with the scientific, transforming both in the process?

2. *Resistance to scientific teachings*

Perhaps it is because of the dominance of science in the modern world and the nature of its teachings that movements sometimes described as "anti-science" have developed. A meeting at the French university of Orsay in June 1970 concluded that science was a religion. Going further, certain scientific explanations like the big bang have sometimes been considered to be myths. In addition, criticisms of science can be used as a justification to reduce public funding of science. Such movements criticizing science have been partly influenced by philosophical debates about the nature of scientific progress. Karl R. Popper asserted that science was a process by which wrong ideas could be disproved or "falsified". In opposition to this point of view, Thomas Kuhn asserted that the science of a particular epoch depended on basic assumptions or "paradigms", which needed a revolution to be overthrown. Historical examples such as the revolution in physics at the start of this century, to be described in chapter 3 of this book, can be used to illustrate Kuhn's ideas. A more extreme point of view is that of Paul Feyerabend in "Against Method" (New Left Books, London 1975) who went much farther, proposing an "anarchist" theory of knowledge, according to which quite irrational methods could be and have been used to produce accepted science.

Another example of "anti-science" movements is the one concerned with ecology. Threats to the environment presented by modern technology are thought in such circles to be due to the nature of science. The French journal "Survivre et Vivre" raised questions of this sort at the beginning of the seventies.

Feminists have stated that science is a result of male domination. A feminine science would be more intuitive. Associated with a corresponding technology, it would not aim to "dominate" nature as does present-day science and technology. However, I must say that the science of my female colleagues, who are relatively numerous in French astronomy, does not appear to be basically different than that of male colleagues.

Science was criticized in the sixties and seventies, especially by sociologists, from a Marxist point of view. The development of science was considered to be a result of society possessing a class structure in which such a development occurred; for example it was stated in the British "Radical Science Journal"(see Bob Young 1977) that "science is social relations". Marxism is no longer in fashion, particularly following the political upheavals in the world which occurred at the end of the eighties and the beginning of the nineties. Sociological criticism of science has, however, continued. "Social constructivists" claim that science is as much the result of the debates between scientists as it is of experiment. However, I must say as a physical scientist that sociologists appear to me to be often not very clear in their thinking.

I support an intermediate position in this book. Science is social relations, but social relations in a much wider framework, in which they are not only relations with other human beings (including those between a scientist and other scientists) and groups of human beings, but also with the world of nature as well as that of pure ideas. Our relations with the latter world dominate, moreover, the ways we use to explain nature.

3. *The aims of this book*

In Chapter 2, I will look at the often forgotten basic assumptions of physics and attempt to show how they fail to explain many fundamental aspects of reality. Various aspects of the history of science are examined in that chapter, which also discusses the paradoxes and contradictions that reliance on the basic assumptions leads to. The discovery of very strange worlds of twentieth century science involving physics and mathematics is described in Chapter 3.

We shall see what are some of the consequences of following the assumptions described in chapter 2, which eliminate consciousness, conscious beings and soul. Consciousness and soul cannot be completely eliminated however and, as will be seen in Chapter 4, there are strong indications that consciousness and soul reappear in certain ways without people being aware of them, especially in contemporary physics. It is in that chapter that new interpretations of certain properties of the world we live in are proposed, involving the presence of different kinds of conscious beings. These interpretations then suggest the possible new directions for research indicated in chapter 5, which may help to lead to a different kind of science. Conclusions are drawn in this last chapter.

It is possible to state the aims of this book in another way. The well known twentieth century philosopher Karl R. Popper, in his book "Objective Knowledge—An Evolutionary Approach" (Clarendon Press Oxford 1972) speaks about three "worlds". The first is the physical world or world of physical states. The second is the world of mental states, while the third world is that of ideas in an objective sense, which can be possible objects of thought. This indicates that the second world is that of subjective experience, while the third world is that of pure ideas, which show themselves in human culture. Each human being must then possess his or her own second world, which will be different from that of any other human. This means that there will not be one second world in this framework, but rather a very large number of second worlds. According to Popper, the first two worlds can directly interact with each other, while the last two can also do so in a similar way. The mind provides an indirect link between the first and third worlds; the ideas of the third world can be made material through technology. Popper states that the third world is a product of human activity. If we try to understand present-day science in the framework of Popper's ideas, we might say that it is a science of the first world obeying the mathematical laws of the third world. This science then eliminates the second world.

The British mathematician Roger Penrose, in his book "Shadows of the Mind" (Oxford University Press 1994) also supports the conception of three worlds. However for him the third world is the eternal world of "Platonic" ideas, which is independent of human beings and includes the unchangeable concepts of mathematics. According to Penrose, the Platonic world gives rise to the physical world, because nature is describable in a mathematical way. The latter gives rise to the mental world, which in turn again gives rise to the Platonic world. His scheme is like that of a serpent chasing its own tail. Roger Penrose is more open-minded than most contemporary scientists and his reputation is not always good in official circles for that reason. In particular he believes that human thinking cannot be reproduced by the kind of computers constructed up to now. However he runs into a barrier, as he still appears to consider reasoning based on the world of physics as fundamental. Apparently he is not able to completely abandon the basic assumptions of present-day science.

Another recent author who talks about the soul is the Estonian astrophysicist Undo Uus ("Blindness of Modern Science", Tartu Observatory, Estonia, 1994). He emphasizes in his book that inner experience cannot be explained by physics.

It is very easy to see that the three worlds correspond to the traditional idea of body, soul and spirit. This threefold scheme was forgotten, if not suppressed, in western culture, in which the dualism of body and soul was taught for a long time, until the soul was then also eliminated by materialism. However the reality of the threefold nature of experience has forced its reappearance in contemporary thought. In this book I endeavor to show that the soul, or second world, is basic; it is present in certain ways also in the two other worlds. It is only when this is taken into account that a fundamentally different kind of science can arise. It is not sufficient to produce some sort of unorthodox physics, as is often done by people trying to prove that something "spiritual" exists, if soul and spirit are left out. The universe has, in fact, a threefold trinitarian nature, revealing itself in several different ways. Such ways of thinking about the Universe will be discussed in this book.

4. *What Do We Mean by Soul?*

In order to progress in the kind of work this book is concerned with, we must be very precise. It is not sufficient to talk vaguely about the soul if we wish to be scientific about it and attempt to lay the foundations of a new science. We must therefore clarify the nature of the world (or rather worlds) of inner experience, or soul, which I shall consider to belong to many sorts of conscious beings. (My approach to this question has been inspired to a great extent by Rudolf Steiner, the Austrian philosopher and spiritual teacher, already mentioned in the preface to this book.) When we examine the worlds of inner experience we find that when they are considered by themselves they also have a threefold character, so that three different aspects can readily be distinguished. These aspects are:

a) Knowledge, which, in the case of the inner experience of human beings, is the result of combining perceptions with concepts through thinking. Both perceptions and concepts are experienced in the inner world. (The fact that knowledge is arrived at in this way was emphasized by Rudolf Steiner in his "The Philosophy of Freedom" (Rudolf Steiner Press 1964). Other quite different ways of arriving at knowledge are conceivable however.

b) The world of feelings, emotions and desires.

c) The ability to act, so as to change the world.

These three aspects can be related to thinking, feeling and willing. The idea that the human soul possesses these three abilities is relatively new. The late eighteenth century German philosopher Johann Nicolaus Tetens in his book "Philosophische Versuch" (1775, re-edited by Verlag von Reuther & Reichard, Berlin 1913) seems to be the first to have proposed that the soul possesses the three basic abilities of feeling, understanding and willing. He defines in this connection the will as the ability to be active, excluding the abilities of representing things (making mental images of them) and thinking. Curiously, one modern philosopher misunderstood Tetens and, when writing about his ideas, said that willing was unimportant for him, which is certainly not the case.

In this book I shall endeavor to show that we can understand the universe as composed of many different kinds of conscious beings, each of which has what we may think of as a kind of soul nature and that many of these beings possess what can be considered as transformations or metamorphoses of the three aspects of the inner world. As a result, relationships between many of the different beings will depend on these three aspects. Let it be emphasized in particular that we shall suppose that the ability to act is a reality and not an illusion, as supposed by many thinkers.

Though one can object that the various schools of psychology have different conceptions about the inner experiences of human beings, we shall see that the three aspects just mentioned and their relationship to conscious beings are not arbitrary, but appear to be very fundamental, especially if we wish to do another type of science. In chapter 2, we shall see that the three aspects or soul abilities are even required to understand the nature of something as basic as time, while the possibility of understanding certain results of twentieth century science through them will be described in chapter 4. In this way, I endeavor to show that a faint gleam of what can become another science in fact already exists. Future work might then be expected to develop these aspects to a much greater extent.

2

The Basic Assumptions of Science and their Limits

1. *What Does Physics Basically Study?*

Let us now see what makes up the world of physics, which is usually considered to be the most basic of the sciences. This world must clearly be derived from what is experienced by human beings, that is from their inner world, which includes perceptions and concepts. In fact physics, as it exists at present, only studies certain kinds of perception, which it associates with a narrow range of concepts. This means that only a small fraction of the world of sense impressions and of inner experience has been used to make up what is considered to be the world of physics, that is, the only world in which the materialist believes.

The perceptions that have been studied since the revolution in ideas in the sixteenth and seventeenth centuries, which led to the birth of modern science, are usually produced by experiments. In experiments situations are especially created, so that certain phenomena can be studied more easily. In fact the will of the experimenter is used in the search for knowledge when experimental methods are applied.

Furthermore, only certain kinds of perceptions are studied. Physics has been more and more monopolized in its development during the last centuries by the study of the interactions between phenomena and measuring instruments, that is, between what is produced by matter according to the conceptions of physics and what is made of matter according to the same conceptions. Direct perceptions of phenomena by human beings, of sensations due to phenomena which are inner experiences, have come to be considered as untrustworthy and therefore are ignored as much as possible. What is trusted is what is considered to be "objective", that is, the interactions between phenomena and instruments, which leads to numerical measurements including, for instance, very small quantities, those which last an extremely short time, and very faint astronomical objects which

8

cannot be directly perceived by humans without the help of instruments. What are studied are therefore the numerically measurable interactions between various phenomena and the instruments used in their examination—as perceived, however, by human beings! It is these processes which are usually considered worthy of investigation in physics. In this way, the role of the human observer is reduced to a minimum.

An astronomical example concerns observations of distant objects without it being possible to perform real experiments on them. In the seventeenth century Galileo looked through a primitive telescope and discovered many things such as the satellites of Jupiter, the phases of Venus (Venus usually appears far from circular and rather looks like the Moon during different phases) and sunspots. The human eye was replaced by the photographic plate about a century ago. More recently, the photographic plate has been replaced by electronic detectors, the use of which makes numerical measurements relatively easy. Such measurements are proportional to the intensities of different kinds of light falling on a particular detector. In that way the human eye has been bypassed; this means that the effect of light on detectors replaces direct experience using the eye.

The restriction in the concepts used by physics is even more striking. The situation is not only that physics studies what can be expressed mathematically, but that in particular it studies what can be expressed in terms of space and the space-like properties of time. This can be understood if we look at Newton's laws of motion as stated in his famous book of 1686, "Philosophiae Naturalis Principia Mathematica", on which mechanics is based. These laws, preceded by Galileo's research in the early seventeenth century, can be considered as being near the beginning of what has become present-day physics. Let us recall Newton's laws, which can be formulated as follows:

1) If no force is applied to it, a body will continue to be at rest or to move in a straight line at a constant velocity.

2) The ratio of the force acting on a body to the rate of change of the body's velocity due to the force, is constant.

3) If one body acts on another the force of action on one is equal and opposite in direction to the force of reaction of the other.

In addition, we have Newton's law of gravitation:

4) Two bodies attract each other with a force proportional to the product of their masses (the masses multiplied by each other) divided by the square of the distance between them.

The second law leads to a definition of mass, because the change in velocity of a body acted on by a given force is inversely proportional to its mass (that is, to 1/mass). Velocity is the rate of change of distance with time, while acceleration is the rate of change of velocity with time. This means that we can say, as is learned in school, that force equals mass times acceleration. In addition let us emphasize that what is involved are measurements of distance and time, with the mass of a body being a constant defined by these laws. Force is also defined with respect to the laws, while what is called "energy" can be similarly defined. In such a framework, distances may in the first instance be defined by what is measured by measuring rods and time by what is measured by clocks, that is, by what are already simple versions of measuring instruments. Let us note that the time measured by a clock is a space-like quantity; it is a number giving the time after (or before) a certain event. In a description of what happens to a body. it needs to be added to the distance and direction of the part of a body where the event occurs, defined with respect to a point of reference.

The definitions of Newton's laws are directly related to the concept of primary and secondary qualities. The former include distance, time, mass, force etc. and belong directly to the world of physics. Other qualities such as colour, taste, smell etc. are supposed to be due only to how the human body perceives phenomena and so to be outside physics. The idea of this separation is in fact quite old. The fifth century BC Greek philosopher Democritus who, together with Leucipus, suggested that matter is made of indivisible atoms moving in empty space, distinguished between the geometrical properties of atoms such as shape, order, position and size on the one hand and non-geometrical qualities on the other. The latter included taste, colour, brightness and darkness, cold and heat, which were all subjective for him and a matter of opinion (see "Les Presocratiques" by Abel Jeanière, Seuil, 1996). Democritus' books have been lost, but similar ideas were described by the first century BC Roman Epicurean poet Lucretius in his poem "De Natura Rerum", which has survived. The origin of such a form of materialism in ancient times, when little was known of physics, is curious; it makes one wonder whether there were not pre-existent materialistic mystery teachings based on some sort of direct inspiration, where materialistic assumptions were taught. Indeed P. Feschotte in "Les Illusionistes" (Editions de l'Aire 1985) considers that materialism may have been an a priori idea.

A similar distinction was made at the beginning of modern science by Galileo in his book "Il Saggiatore", where he distinguished between what is measurable and what is not measurable, such as smell and taste. He insisted that nature is written in mathematical language and described a scientific method for testing hypotheses. The seventeenth century English philosopher John Locke separated primary qualities such as size, shape and density from secondary qualities like colour, taste and smell.

Another aspect of Newton's laws, and thus of classical mechanics, needs to be emphasized. The second law describes rates of change. If one knows the forces present, including those discovered after the time of Newton, and the positions of all bodies including those of each of their constituent parts at a certain time, it is possible to determine their future positions by adding up the rates of change at each time, or by what mathematicians call integration. The mathematics involved in such calculations due to Newton and Leibniz, developed enormously from Newton onwards. In this way the universe came to appear to be completely deterministic—though as we shall see in the next chapter, even Newton's laws do not exclude the unpredictable.

It should also be emphasized that laws like those of Newton encourage explanations of physical phenomena in terms of what happens to the smallest constituents of bodies. Each of these constituents, which were earlier considered to be atoms and later particles, can be expected to be acted upon by different forces. What is observed should then be the result of adding what happens to each constituent. According to this very basic conception, our large-scale world is explained by the very small.

2. The Further Development of Classical Physics, Chemistry and Biology

The later development of classical physics and chemistry involved, among other things, the study of phenomena unknown at the time of Newton and in particular the discovery and investigation of forces other than gravitation, without however changing the basic hypotheses mentioned in the last section. This type of thinking also came to dominate biology. Such developments became particularly important in the nineteenth century. In many ways, twentieth century biology is also part of classical science and will for this reason be considered together with physics in this section. I will now give some examples of results which were very important for the way modern science was to evolve after the laying of the foundations of mechanics.

The study of electricity and magnetism made great progress in the nineteenth century. The forces of electricity and magnetism were found to be closely connected. The Danish physicist Oersted discovered the magnetic effect of an electric current in 1820. Following the work of many other physicists such as Ampère and Faraday, Maxwell produced a general theory of electromagnetism. Light was explained as consisting of electromagnetic waves having small wavelengths which are visible to the human eye, the wavelength of any sort of wave being the distance between two successive peaks of the waves. Other kinds of electromagnetic wave, such as the radio, x and "gamma" rays, were later found to have different wavelengths from those of visible light. In fact there had previously been a clash between those who considered light as consisting of particles, an idea favored by Newton, and those who considered light to be due to waves. The latter conception seemed in the nineteenth century to have been proven by the work of Young and Fresnel. The clash between these ways of explaining light was not, however, concluded by Maxwell's electromagnetic theory, as we shall see in the next chapter.

At the beginning of the nineteenth century, Dalton explained the processes of chemistry by interactions between atoms. Atoms of different fundamental substances, called chemical elements, each having different properties, combine to produce what are now called "molecules" containing several atoms. Each substance obtained by combining different elements is called a compound, and has, according to this conception, its characteristic molecule, while substances containing different molecules have different chemical properties.

Another example of the development of physics is thermodynamics or, in more popular language, the physics of heat. Changes in physical properties of bodies, such as expansion due to increasing temperature (a measurement of expansion is clearly one of changing length or distance) led to a definition of temperature. The pressure of a gas, that is the force per unit area exerted by the gas on a wall, was found to depend on temperature. According to the very successful kinetic theory of gases, this was explained by the fact that a gas consisted of particles (in fact molecules) which struck the wall, each molecule having an average velocity proportional to the square root of the temperature above absolute zero. Heat was generally explained by the disordered motions of molecules. In this way the human experience of heat became explained as only being due to an average velocity of small particles.

It was through the explanation of the phenomena of heat that the concepts of probability and statistics entered physics. It was far too complicated to calculate the motion of every molecule. Each molecule was therefore assigned a probability of having certain properties; it was the mean properties of the molecules of a substance which were then to be considered. Clearly this did not contradict Newtonian determinism; it was, however, only necessary to calculate mean properties in order to explain the large-scale structure of the world.

It should be noted at this point that the concept of energy came to be considered fundamental in physics and chemistry. For example, energy was found to be contained in electricity, magnetism and light, to be necessary for or produced by chemical reactions, and to be present in heat. The role of energy is in fact that of a sort of "ability to act" of a physical system. The total amount of energy in any system which is isolated from the rest of the Universe was found to be constant, this being called the "principle of the conservation of energy".

The contribution of Darwin in the nineteenth century was fundamental to the history of biology. The theory of natural selection which he proposed and which was developed by those who followed him, states that when random variations (now called "mutations") in the abilities of living organisms occur, those organisms which are more able to survive will survive. In this way evolution can be produced, because new kinds of organisms will appear and many sorts which are less able to survive will be eliminated. Nothing outside the laws of physics, whose basic nature we have already looked at, is needed for this kind of evolution to occur.

Similarly, Mendel founded genetics. He looked for easily identifiable characteristics of peas and measured the proportions of different sorts of descendents, produced when peas with different characteristics were crossed. He found certain laws that these proportions obeyed and in this way he found normally constant factors in heredity (which are only modifiable by mutations), later to be called genes. A Gene can be thought of as being a kind of "atom" of heredity. This was to lead to the present ideology about how genes dominate the living organism to the exclusion of everything else. The nature of genes has been elucidated in the twentieth century and appears to be based on the chemical properties of certain very large molecules, that is, the molecules of a substance called DNA. Modification of the DNA molecules of living organisms has now become possible, such manipulations being called "genetic engineering". However, as emphasized by Craig Holdrege in "A Question of Genes. Understanding Life in Context" (Floris Books, and Lindesfarne Press with another title, 1996), all the phenomena of life are far from being explained by genes. This means that life cannot be put into such a straight jacket.

It was also in the nineteenth century that Karl Marx claimed that he was able to explain the nature of human society materialistically, in a way which is somewhat similar to the explanation of evolution by Darwin. Everything, according to him, is based in the last resort on economics, this being particularly clear in that part of Marxism called "historical materialism". This is clearly very one-sided, though the effects of the economy on human society and the way people think should not be underestimated. In any case, any explanation of economics only by the laws of physics is, to put it very mildly, highly doubtful. Marxism became a very powerful force in politics and in the twentieth century it was to be used as a state ideology. It was this state ideology, justifying extreme forms of dictatorship, that later failed.

One thing which is already striking in this description of the history of science is the way in which science, and especially physics, "succeeded" in spite of the limiting nature of its basic assumptions. This apparent success continued into this century with the rise of modern physics, including relativity and quantum theory, to be discussed in the next chapter.

The physical definitions of space and time have become more subtle and very far from what is suggested by normal human experience; but space and time still remain basic. Such a success would appear to show the real presence of a world of the physical sciences, or material world, which is not the same as the world of human experience. The nature and possible interpretation of this world will become clearer later in this book.

3. Critiques of the Basic Assumptions of Physics

The most fundamental criticisms of Newtonian physics were made by the great German poet, writer, thinker and scientist Goethe, who lived at the end of the eighteenth century and the beginning of the nineteenth. His scientific work on the nature of colour and the development of plants is particularly noteworthy.

In perceiving phenomena, he sought to directly perceive the ideas behind them in a rigorous way, without making hypotheses. In this way he perceived not only the principles of the way plants change in their development, or what he called their "metamorphosis", but he also perceived the idea of an "archetypal plant", containing the whole nature of plants. When we examine Goethe's way of looking at the world, we are in a completely different conceptual framework than that of classical (pre-twentieth century) physics. For him the same laws appeared in nature as in art; artistic creation was the same creation as natural creation at a higher level. In addition the senses of human beings were, according to Goethe, the greatest and most precise of physical apparatus; the fact that physics had detached experience from man was the worst of misfortunes.

Goethe's theory of colours radically challenged what was believed by physicists. According to Newton, white light is made of a mixture of lights of different colours. He found that a narrow beam of white light passing through a prism was split into beams with different colours, each of these lights being arranged in what is called a spectrum. Red coloured light in a spectrum is followed by lights having colours of orange, yellow, green, blue indigo and violet. Pupils in physics are still taught this, as they are also taught that the quality of colour as perceived by human beings is something outside physics, only associated with sensory perception. Moreover, according to the nineteenth century explanation of light as being composed of those electromagnetic waves which are visible to the human eye, the sequence of colours in a spectrum is a sequence of electromagnetic waves having different wavelengths, the wavelength of "pure" red light being larger than that of "pure" light having other colours. Goethe noticed, however, that if one looks directly through a prism colours appear only at the boundaries of white and dark surfaces; in fact one sees only part of a spectrum at a boundary. A complete spectrum is perceived when a small white spot on a black surface is observed through a prism, while a dark spot on a white surface produces a sort of "negative spectrum", consisting of the opposite, or what are called complementary colours to those of the normal spectrum. In fact this does not contradict Newtonian theory, for according to that theory each of the points on the white surface will produce a spectrum which is shifted with respect to the spectra of other points; the superposition of all the spectra will mix each of them to produce white light except at the boundaries of the white surface.

But Newton's ideas appear, according to the results of the last experiment, to be artificial and not directly deducible from human experience. For Goethe colour really played a role in the world and was produced by the interaction at the boundaries of light with darkness, which also had a real existence. This interaction was mediated by substances such as that of the prism. It may be noted that the idea that colour is produced by the interaction of light with darkness is much older; in his book "Timaeus", Plato already suggested that different colours were produced by mixtures of black, white, red and what is bright. Aristotle went further, stating that white seen through a black screen appears to be red.

Goethe's conceptions have not played a major role in official science. Indeed, the situation at present is even worse than it was during Goethe's time. There is a tendency, especially in astrophysics, to wish to calculate and deduce everything from supposedly rigorous theoretical models, using the very powerful computers now available. However, such models can still contain highly dubious assumptions, even from the point of view of official physics, while they can also neglect certain observations. This is clearly the opposite of the Goethean approach! In spite of this, a few scientists have tried to base their ideas directly on observations. An example is the well known Swiss astrophysicist Fritz Zwicky, who died in 1974. In a book "Morphological Astronomy (Springer 1957), he describes his unorthodox way of doing astrophysics. I tried to be at least "slightly Goethean" at the start of the work leading to my doctoral thesis, as well as in some later research. In this work I looked more closely at what observations can tell an astronomer than is usually done. This led me in my doctoral work to propose explanations and support models which were "out of fashion"; the fashion changed about a decade later. In any case, this type of "semi-Goethean" method does not challenge the basic assumptions of physics. More radical Goethean challenges to present-day science have not yet made much of an impression either, perhaps because of their inability, unlike that of official science in general and even nineteenth century physics, to explain from simple assumptions many different phenomena or at least their space-like aspects. As already mentioned, classical physics was amazingly successful. It appears to me in view of this that even more radical approaches are required.

Another aspect may be mentioned concerning the nature of space, on which physics is based. It has three dimensions and so can be measured in any three perpendicular directions from any point. This is the basis of what are called "Cartesian coordinates" in geometry and their role in the calculations of physics can be very abstract. Rudolf Steiner pointed out in his lecture cycle "The Origins of Natural Science" (Rudolf Steiner Press and Anthroposophic Press 1985), that there are basically three fundamental perpendicular directions in human experience; or, to be more precise, in the experience of the human body. These are the directions from back to front, from left to right and from down to up. It is those directions and not any three possible perpendicular directions which are important for Man. In the world of physics a particular physical system might have preferred directions, but in general calculations can be made using any three perpendicular directions; also from this point of view physics, in becoming abstract, has become disconnected from what is directly perceived by humans.

4. Arguments about the Nature of Consciousness

At this point I shall insert descriptions of certain problems which excite much interest at the present time. These problems show the contradictions inherent in the foundations of classical science as described at the beginning of this chapter, and suggest that another kind of science is needed. They can be looked at in many ways without invoking twentieth century physics—which will be considered later in this book.

There has been much discussion in recent years about the causes of consciousness and particularly thinking, and about how to explain them using the current materialistic ideas concerning the nature of the world. Scientific results about the brain and nervous system are generally used as the basis for these discussions. It is often even asserted that all the phenomena of consciousness and thinking are reproducible by calculating machines and computers. In view of what has been already stated in this chapter about the physics on which materialism is based, such discussions seem extremely curious. Consciousness has been eliminated from physics and now certain scientists want to deduce it from this physics without consciousness and the science of computing in order to explain human experience and also that of higher animals!

It is not my intention to describe in detail these debates, which appear to me to be rather futile. More can be found in the already mentioned books "Blindness of Modern Science" by U. Uus and "Shadows of the Mind" by Roger Penrose. Let us add "The Emperor's New Mind" by Roger Penrose (Oxford University Press 1989); a debate about artificial intelligence contained in two articles in the January 1990 issue of "Scientific American" by the Berkeley, California philosopher John R. Searle and the Churchlands, as well as an article by Searle which appeared in the May 1996 issue of the French general scientific journal "La Recherche".

Discussions are generally centered on the question of whether or not consciousness and thinking can become (or are) properties of computers. The question is not only scientific, as there are large economic interests involved in the development of more powerful computers and robots. If such machines can be made more able to imitate the processes of human thinking, using what has been discovered about the nervous system, the computer industry will be able to make large profits. Roger Penrose in "Shadows of the Mind" describes four points of view on these questions:

A. Strong or hard artificial intelligence: All thinking is computation; even feelings of conscious awareness are evoked by the carrying out of appropriate computations. Strong artificial intelligence can also be considered (as in Searle's article) as proposing that the mind is only the "software" of the brain, the latter being the "hardware" of what enables a human being to think; a human must then be considered to be a sort of computer. In fact the software of a computer is its system of programming and its hardware its physical structure.

B. Weak or soft artificial intelligence: Awareness is a feature of the brain's physical action and although any physical action can be simulated computationally, computational simulation cannot by itself evoke awareness.

C. Appropriate physical action of the brain evokes awareness, but this physical action cannot even be properly simulated computationally.

D. Awareness cannot be explained by physical, computational, or any other scientific terms.

Roger Penrose strongly argues for C, because it has been proved that computers of the kind which exist at present cannot perform all mathematical proofs that can be performed by mathematicians, while D is for him the point of view of the mystic and outside science. As far as the arguments given against A and B by Penrose are concerned, it must be emphasized that the performance of all mathematical proofs is impossible for any machine based on the processes of classical physics. Roger Penrose tries to overcome this dilemma by supposing that the nervous system works according to the principles of quantum physics. The model proposed by Roger Penrose is, however, quite speculative. His opinions on this subject have made him rather unpopular in certain official circles, but in fact he does not escape from materialism. Here I shall try to show that there are other possibilities similar to D, which are not "outside" science.

In the 1996 article Searle rejects A, as he has done in previous articles. Minds have contents and, unlike computers, do not only handle symbols according to certain rules. He then describes three approaches in a way different from Penrose. Searle does not accept Penrose's reasoning, which he strongly criticizes, while the studies of the nervous system by two other authors (Crick and Edelman) are considered helpful. Searle admits, however, that the qualities of mental experience or "qualia" as they are called, cannot be explained in such ways. In fact, qualia belong to Popper's second world and though researches on the brain and nervous system indicate that certain mental events are to quite a large extent correlated with events in these parts of the human body and therefore with phenomena of the world of physics, one cannot really expect to explain qualia by using present day science!

5. *Modern Astronomy, Cosmology and the Anthropic Principle*

The contradictions of present-day physics are also clear in discussions about what is called the "Anthropic Principle", This principle was first stated by Brandon Carter in respect to cosmology, that is, the science of the large scale structure and evolution of the universe. To be studied scientifically, the universe had to be able to produce intelligent creatures, that is, humans; this means that it had to possess certain properties. In this way the existence of these properties is explained by the presence of intelligent beings!

This principle appears at first sight to contradict another basic principle used in the study of the universe for more than four centuries, the "Copernican principle". This principle, according to Konrad Rudnicki in "The Cosmological Principles" (Jagellonian University, Krakow 1995), asserts that the universe observed from any planet looks much the same. Therefore the planet on which Man lives has no special significance compared with that of other planets. The Copernican principle can, according to Rudnicki, be extended to obtain a more general principle, which he calls the "generalized Copernican Principle". According to this last principle, the universe observed from every point and in every direction looks much the same. This principle, if true, leads to the conclusion that the home of the conscious being Man not only has no special significance compared with that of other planets, but also has no significance with respect to that of any other point in the whole universe. Indeed one can argue without invoking the generalized Copernican principle that it is "unlikely" that our home is at such a significant point. Such a principle can clearly be connected with the rejection of an important role for consciousness by present-day science, though it is of course not equivalent, as consciousness and intelligence can be conceived as existing in many different places in space as well as on earth.

Before saying more about the anthropic principle, it is necessary to summarize what astronomers consider to be the nature of the universe. The planets revolve around the Sun, which appears to be a fairly normal star. The distances of the stars measured by different methods are fairly consistent and the faintness of other stars compared with the Sun can be understood by their much greater distances. Stars are considered to consist of hot opaque gases, which are hotter in their interiors than at the surfaces, from which the light seen by observers directly comes. The energy of almost all stars is thought to be produced by processes known to nuclear physicists in their extremely hot interiors, and the way stars change or evolve during their existence is predicted using the laws of physics. The

Sun seems to belong to a system called the Galaxy, which contains hundreds of thousands of millions of stars. Many other galaxies exist, in some of which the brightest individual stars can be detected using modern instruments. The American astronomer Edwin Hubble argued convincingly in 1925 that certain objects seen in the sky were not in our galaxy. There are indications believed by almost all astronomers (there are a few exceptions) that the galaxies are moving away from each other. Galaxies further away from our galaxy have, according to these indications, a higher velocity relative to our galaxy; it is impossible to observe parts of the universe beyond a "horizon", moving away from us at speeds greater than that of light. In this way the universe is thought to be expanding; it must then have been very dense, very hot and very much smaller than it is now in the very early stages of its development. The original state is often called the "big bang", as the idea of it resembles that of an explosion. Since light takes a long time to come from far-away objects; the universe must have been much younger when the light from the farthest visible astronomical objects was emitted.

Let us consider in a little more detail, though still schematically, what is thought to have been necessary to produce human beings. According to the conceptions of cosmologists, the constants of physics, which could have been different originally, became fixed at a very early stage in the expansion of the universe, and the different fundamental forces of physics became separate forces. Then, according to these conceptions, as the universe cooled a few chemical elements were formed by certain processes of nuclear physics, but most gas remained in the form of hydrogen. The universe became transparent later, which it had not been before. The diffuse gas, which filled the universe at that time, then condensed in galaxies and especially into stars inside galaxies. Most chemical elements, including carbon nitrogen and oxygen, which play a primordial role in life on earth, are thought to have been manufactured in the interiors of stars by processes of nuclear physics, which are different from those of the very early universe. These elements were then ejected with other material from the stars in which they were made into the space between stars and often became incorporated into other stars, which condensed later. In particular they became incorporated into the Sun and planets. Life, which is considered to be a property of very large, complex molecules, probably needing to contain a very large number of carbon atoms, could then be created in the conditions existing on earth as a result of chemical processes. A very long period of evolution then would have been necessary to produce intelligent beings like humans with large brains, by the random process of Darwin's natural selection.

For such a development to be possible, the universe must be very old; its age is in fact estimated as being of the order of fifteen thousand million years, though it should be noted that there is some disagreement about the exact value of its age. This means that it must not have had properties which would have made it start to contract soon after beginning its expansion. An expanding universe which is very old must clearly be also enormous, as it would have needed a long time to expand. The constants of nuclear physics must also have been suitable for the formation of the chemical elements necessary for life in roughly the right proportions. In such ways the similarities of the ratios of certain physical constants which have been discovered, might be explainable. When such conditions are taken together, it is clear that even according to such materialistic ideas, we cannot live in a random universe.

Discussions about the anthropic principle are given in a number of places, such as "The Anthropic Cosmological Principle" by Barrow and Tippler (Clarendon Press 1986) and the already mentioned "The Cosmological Principles" by Rudnicki. In fact there are several forms of this principle, which are stated in different ways:

The weak anthropic principle: This asserts that human beings must live in a universe which could have produced them or, as stated by Rudnicki, "the physical properties of the observable part of the universe have to be taken as a logical conclusion from the premise that the human being observes it". Barrow and Tipler define the principle without directly referring to human beings as "The observed values of all physical and cosmological quantities are not equally probable, but they take on values restricted by the requirement that there exist sites where carbon based life can evolve and by the requirement that the universe be old enough for it to have already done so."

The strong anthropic principle: This asserts, according to Barrow and Tipler, that "The universe must have those properties which allow life to develop within it at some stage of its history". Rudnicki's different definition of this principle is that the "physical properties of the universe have to be taken as a logical conclusion from the premise that real observers exist in some parts of the universe's space-time".

The final anthropic principle: Suggested by Barrow and Tipler, this states that "Intelligent information processing must come into existence in the universe and once it comes into existence, it will never die out". This principle is related to their very daring, highly speculative hypotheses about how the universe containing intelligent beings able to process information should end—that every civilization produced by such beings is able to attain a point when, as well as defending itself successfully from various perils, it is able to create/construct more intelligent and more resistant beings (in fact, robots able to survive in very extreme physical conditions) than those of the original civilization. The ultimate descendents of such civilizations may then possibly encounter those of other civilizations; in that case the civilizations will merge. Life will obtain unlimited knowledge and gain control of all matter and forces, take over the universe, which in a very materialistic way will (perhaps must?) then have a "happy end".

It is clear that the first of these forms of the anthropic principle is the least speculative and the last the most.

Specialists are usually very uncomfortable when they speak about the anthropic principle, because it violates to a smaller or greater degree their basic assumptions, according to which consciousness should be unimportant. The weak anthropic principle cannot be denied, though one can always say that the conditions necessary for all forms of intelligent life are not really known. Even according to materialistic assumptions, other sorts of intelligent life than that which exists on earth might in principle be possible in this as well as in another sort of universe. If this possibility is denied, there is another sort of escape for the convinced materialist. A very large number of different universes could exist, so that humans only live in one of the very few in which intelligent life is possible. Other universes could exist (or have existed) earlier or later in time than our present one, be very far beyond the "horizon" and so unobservable, or be permitted by a certain interpretation of what is called "quantum physics". Quantum physics and its interpretations will be discussed in the next chapter of this book. In this way, Barrow and Tippler "rescue" their materialistic ideology.

It seems clear from this discussion that it is rather difficult to suppress consciousness in any conception which tries to explain the whole universe. If you try to do so you tend to run into paradoxes.

6. *The Nature of Time*

As we saw at the beginning of this chapter, present-day physics is based on space and the spacelike aspects of time. It is easy to show that this concept does not even take into account all the aspects of time, which have puzzled thinkers for a very long time (no pun intended!).

The major problem is that time has not only spacelike properties, but in addition it consists of past, present and future. The present is not stationary, but "slides" from past to future. For instance, the second century AD Roman philosopher emperor Marcus Aurelus compared time with a flowing river. In this he may have been inspired by the 5th century BC Greek philosopher Heraclitus for whom change was fundamental and who said that one never entered the same river twice. St. Augustine (5th century AD) was extremely concerned about time in his "Confessions", where he prays to God to enlighten him. He wonders whether the past and future can, like the present, also exist—and about the relation between time and movement. St Augustine concludes that time only flows for the soul. This interest in time was continued by the early twentieth century philosopher Bergson. We should also mention Einstein, who stated that the problem of time worried him seriously; the present, essentially different from past and future, meant something special for Man, but that this important difference did not and could not occur in physics. Such problems now interest a number of contemporary physicists. In this connection we can mention the book "Now, Time and Quantum Mechanics" (editors Michel Bitbol and Eva Ruhnau, Frontières, 1994) and a small popular book in French "Le Temps" by Etienne Klein (collection "Dominos", Flammarion, 1995). Let us now look at these questions in more detail.

In the world of classical deterministic physics of Newton's laws the passage of time is an illusion, because the future, like the past and present, already exists in a certain way. This is easily seen, as it is in principle possible according to these laws to completely calculate the future, which then cannot bring anything new into existence. Indeed the whole of time is in such a situation similar to the human perception of the past.

The first property that distinguishes time from space is that it has a direction or "arrow". This becomes clear if a film is run backwards and "impossible" things happen. The pieces of a broken cup are put together on the floor and then rise spontaneously to the top of a table. An undamaged house emerges from a fire, human corpses rise from the dead, walk and if one waits long enough become babies, which then enter their mother's womb. In a less spectacular example, if time could be run backwards, temperature differences would spontaneously emerge in regions where the temperature was uniform. However, according to Newton's laws of motion when they are applied without any additional law, reversing the motions of all particles as in the film run backwards should not produce situations which are radically different from those in which motions are not reversed. In fact, in all such phenomena a basic law of thermodynamics manifests itself, that is, that in an isolated system a quantity called "entropy" must increase with time. This quantity measures the disorder of a system. The positions and motions of its molecules should, because of the mathematics of statistics, evolve into more probable states, probability being defined by the statistical considerations mentioned in section 2 of this chapter. Nevertheless, this situation appears paradoxical because according to a famous theorem of the French mathematician Poincaré based on classical physics; if one waits long enough events which are almost identical with previous ones will occur in a finite isolated system. In this way past events should eventually be repeated.

It is of course doubtful whether the whole universe is a finite, isolated system. The arrow of time in any case indicates that the universe has not only an already existing past, but that it also has a future. There is a direction in its evolution. In fact cosmologists are able to explain the increase in entropy and the arrow in a quite materialistic way starting from what is supposed to have happened following the origin of the Universe in a "big bang". An expanding universe of the sort believed in by cosmologists has a beginning which is quite different from late stages in its development. The "arrow" of time can still be integrated into present-day physics, without changing fundamental ideas.

The most difficult problem occurs if the present is included in the description of time, because it would then appear to be "outside physics". It is here that a science which does not take account of soul qualities has some of its greatest difficulties. These difficulties can be overcome if we examine the human experience of time, taking into account the three soul abilities mentioned in Chapter 1. A human being is in the present, but from this vantage point the future, present and past are experienced in quite different ways. He or she is able to act, that is, to use his or her volition in the present to influence future events, which he or she cannot predict with certainty. This unpredictability has been confirmed by twentieth century science, as we shall see in the next chapter. As a result, the future appears to be "dark" for the human being. The possibility of action dies when the future becomes the present; he or she will then tend to have the strongest feelings about the results of the actions performed, such as feelings of satisfaction, or dissatisfaction, or pleasure or remorse. It is even possible that to have such strong feelings about the future or the past is mentally unhealthy. A human being can see the past, have knowledge of it and so reflect on it with the greatest amount of clarity using his thinking ability; the past however, unlike the future, is dead. What is known about the past is used to think about the future, in so far as this is possible. When we think about the nature of the human experience of time, we can see that we are not making an arbitrary division of the soul when we refer to the three soul abilities. They appear to be basic in the nature of the world. Time is fundamentally a soul experience. This true nature of time has not, until now, been taken into account by physics. My manner of looking at time is inspired by Rudolf Steiner, especially a lecture he gave on November 27, 1920, published as part of a series called "The Bridge Between the Universal Spiritual and the Physical Constitution of Man. Freedom and Love. Isis Sophia" (volume 202 of his collected works), in which he relates time to the three soul abilities.

It is also interesting to note that the English language relates willing to the future. In the future tense of verbs, the word "will" is used. For example, the reader of this book will, after reading this section of chapter 2, perhaps eat a meal or fall asleep!

This soul nature of time is also connected with another aspect of Rudolf Steiner's teachings. In his book "How to Know Higher Worlds", he describes a path of spiritual development which leads to the possibility of having perceptions of the spiritual nature of higher invisible worlds and the attainment of spiritual knowledge. He contends that his statements can be verified by using these methods. At a certain stage of this development, the soul abilities (thinking, feeling, willing) become independent of each other, which results in great dangers if the true self is not strong enough to master such a situation. This can be compared to the fact that under the influence of certain drugs human perception of the order of events in time is disturbed and a kind of temporary madness can set in. This madness would, in view of our considerations about time, appear to be connected with a separation of the three soul abilities in what is clearly a very dangerous and dubious way of having experiences of higher worlds.

One other aspect must now be emphasized. The human way of looking at time can only be true and represent reality if the universe is not completely predictable, that is, if acts of will can influence future events. As we shall see, twentieth century science has in fact found that the world is not completely predictable. There is then at least a possibility of will being able to act meaningfully. This need not only involve human will; it is conceivable that the wills of other beings also act to influence events in such a partially unpredictable world. It is with this kind of question, combined with how a new science can be based on the soul abilities, that we shall be concerned to a large extent in the rest of this book.

3

Twentieth Century Upheavals in Science

1. *The general nature of the upheavals*

In this chapter we shall look at certain major transformations which violently shook the foundations of science in the first third of the twentieth century and which, unlike other already mentioned twentieth century discoveries, made it quite different from previous science. When we examine what happened we can see that a threshold was crossed into something new. In fact, the path followed by the development of science since the seventeenth century, according to the basic principles described in the previous chapter, were gradually to lead it into a very strange world. The world beyond this threshold can teach us a lot from a philosophical point of view and particularly about the possibility of putting soul into science, as will be seen in the later chapters. It is, however, first necessary to understand what happened and for this it is useful to have at least a "taste" of the scientific reasoning involved, which led to the transformation of science. In this part of the book I shall give a fairly non-technical description of the main aspects of the discoveries made and of some of the debates about their significance.

A very important aspect of what was found is that not all things which happen are predictable. This, unlike what was thought previously, is now known to be true even for certain phenomena in the everyday world inhabited by human beings. It is also true in a different rather strange way in the physics of the very small, where what is called quantum physics must be used and, as we shall see, it is even true in mathematics! In addition, the conceptions about the natures of space and time were radically transformed, firstly by the theory of relativity and then more radically by quantum physics, though physics retained its fundamental assumptions based on space and the space-like aspects of time.

2. *The space-time of relativity*

Let us start with a major change in scientific ideas, which revolutionized thinking about space and time, but which did not involve the unpredictable.

For Newton space was something absolute, separate from the bodies which moved in it and for which his laws of motion were true. Similarly, time was universal in classical physics. It was these ideas which were overthrown by the special and general theories of relativity. There is ambiguity in the classical idea of an absolute space and classical physics in fact contains a sort of "relativity". When no force is acting, bodies can, according to Newton's laws, continue in uniform motion with respect to each other. No one uniform motion is to be preferred to another, as Galileo already knew. If two bodies are moving in the same direction with velocities v_1 and v_2, the velocity of one with respect to the other is v_2-v_1. The length of a rigid rod and the speed of a clock used to measure lengths and intervals of time, are then supposed to be constant, as are the velocities derived from the lengths and time intervals.

In a classical framework, it was difficult to understand how light could travel through space. In the nineteenth century Space was thought to be filled by a real substance called "ether", in which light waves were thought to travel. In such a situation one might imagine that absolute velocities could be defined and measured with respect to the ether. The earth should then move through it and the effects of such a motion of the earth with respect to the ether should be measurable. Several experiments were made to measure the earth's motion, including in particular those of Michelson in 1881 and of Michelson and Morley in 1887. Because of this motion, light should have taken slightly different times to travel in different directions. No effect was ever seen except in the work of Miller, which is sometimes quoted by those who challenge the orthodoxy of present-day physics. His effects are, however, generally considered by physicists to be smaller than the errors.

Though physical effects can be proposed to explain the failure to detect the motion of the earth through the ether, Einstein realized that the reason was something more basic. His special theory of relativity of 1905 applied to bodies moving uniformly with respect to each other. For all such bodies the velocity of light always had to be constant, whereas in Newtonian theory no state of uniform motion was to be preferred to any other. The result of such assumptions is that the measurements of space and time are not the same for two bodies in different places having different speeds. When two bodies are moving with respect to each other, each of the bodies will always appear shortened in the direction of its motion, if seen by an observer moving together with the other body. When its speed relative to the other body equals the speed of light, its observed length approaches zero. In the same way a clock on one of the bodies will appear to the observer moving with the second body to be slower than its own clock. If it was moving with the speed of light relative to the other body, the first clock would appear to be stationary, its time having stopped. The mass of a moving body will also appear to increase when it is measured by an observer moving with another body, becoming infinite when the relative speeds are equal to the speed of light. In fact, special relativity leads to the conclusion that no body can be accelerated beyond the speed of light.

Consideration of the fact that the apparent mass of a body varies with velocity leads to the conclusion that mass and energy are equivalent. The relation is given by Einstein's famous equation

$$E = mc^2$$

where E is the energy, m the mass and c the speed of light. What is particularly striking is that this equation is also true for the "rest mass" of the body, that is, the mass it appears to have when it is not moving with respect to an observer. This relation is confirmed by experiments and became famous because of its application in the tremendous energy released by nuclear weapons, which convert a significant proportion of their mass into energy.

Special relativity can also be understood geometrically in the framework of the 4 dimensional "space-time" of Minkowski. This can be defined by an observer with respect to whom other observers are moving as containing 3 dimensions of space and one which is derived from the observer's time (time multiplied by the speed of light and by the "imaginary" square root of -1). The path of a particle in this space is called a "world line". "Distances" in this space divided by the speed of light are then equal to the time measured by a clock near a particular particle, moving with it. It would not be easy for such an observer moving with a particle to decide at what time a distant event occurs. One way would be for him to send out a light signal which is immediately reflected if an event occurs far away, the event being in this way detected. The time of the event could be defined as half-way in time between the sending out and the return of the signal. However, another observer would not measure the same time for the event—even the order in time of two different far away events could be reversed for the two observers. Their experiences in time could be completely different. Minkowski summarized the situation as it appeared to him, by stating that space and time by themselves were doomed to fade away into mere shadows, and only a kind of union of the two would preserve an independent reality. A very clear non-technical description of the apparent paradoxes of special relativity is given in the already mentioned book by Roger Penrose: "The Emperor's New Mind" (pages 191–201).

In 1915 Einstein published an extension of his theory to accelerated motion and gravitation, which is called "general relativity". In this theory space and time are found to be curved. This can be easily be understood if we consider two dimensional analogies. The geometry on the surface of a sphere, for example, will not be the same as that on a flat surface. Straight lines on the flat surface are replaced by parts of great circles on the surface of the sphere which, like straight lines on a flat surface, give the shortest distance between two points. This means that an elastic stretched on the surface of a sphere will lie on a great circle. For a more general surface similar curves are called "geodesics". A gravitational field was found by Einstein to distort space-time and a particle acted on by a gravitational field followed the geodesic. This theory led to a simple explanation of the fact that the mass in Newton's second law of motion can be defined as being the same as that in his law of gravitation (after correcting for the additional effects of relativity).

A number of experimental tests and astronomical observations have confirmed general relativity. For instance, the motion of the planet Mercury can only be fully explained by taking account of general relativity. The planet does not only go round in its orbit, but the elliptical orbit itself also revolves at a rate not predictable by Newton's theory. The bending of light rays passing close to massive objects like the Sun, is also a famous prediction. It is now difficult to challenge general relativity, which is very much used, in particular by those who study the large scale structure of the universe.

However, what must be emphasized is that in spite of the successes of relativity, absolute motion and absolute time have in a certain way returned! The absolute time, a "cosmological" time since the big bang, when the universe is thought to have begun to expand, can be defined theoretically. It is assumed (this not being contradicted by observation), that the distribution of matter in the universe when "averaged" over very large distances, is uniform, having a constant density at a given time—time being defined as what is measured by clocks moving with the mean distribution of matter—and will flow "on average" at the same rate everywhere. As the universe expands, the mean density will decrease while the absolute time increases.

What is thought by most specialists to be absolute motion can be measured from observations. It is believed to be the motion relative to weak diffuse electromagnetic radiation observed to be coming from all directions in the sky. It is invisible to the eye, its wavelengths being much larger than those of visible light. This radiation is thought to be a remnant of radiation emitted everywhere when the universe, which was then much hotter than now, was becoming transparent some time after the big bang (see section 5 of the last chapter).

The subsequent continuing expansion of the Universe would then have considerably increased the wavelength of the radiation. Almost equal amounts of this radiation come from all directions, this property being called "isotropy". The first deviation from isotropy is usually explained by specialists as due to the motion of the observer relative to the radiation, though this interpretation is not 100 per cent certain. Because of an effect in physics called the Doppler effect, motion towards a source of waves decreases their wavelength; motion moving away from the source increases the wavelength. An absolute motion of the solar system very close to 370km/sec. has been measured from observations of this radiation using the COBE satellite.

A number of misconceptions have arisen concerning relativity. This is perhaps because in a certain way the Minkowski space-time of special relativity and the curved space-time of general relativity have replaced Newton's absolute space and universal time. Time, or more precisely its space-like aspects, has become a kind of space, being indeed indistinguishable from the latter. In this way the study only of the space-like aspects of time has led to a sort of logical conclusion, the fusion of space and time. According to this way of thinking, the world line of a particle appears to be eternal, as it is independent of time. In part of a radio programme I heard about science and religion, it was even claimed, from what I remember, that this was evidence for the existence of God.... In fact each observer will experience time in his or her own way, this being different from that of other observers. The time of a distant event, which could be due to an action of a living organism (or, as we shall see according to the ideas developed later in this book, an act of its willing), will appear for a particular observer to be the time when a result of such an act is perceived, or a signal is received from the distant event. If the observer is human he or she will still experience time, as described in the last chapter. The actions of other living organisms perceived at a certain time can still influence what happens at later times, but cannot influence previous events perceived by the observer, as relativity does not contradict the ordinary ideas of cause and effect. Relativity, unlike the other twentieth century developments now to be discussed, does not involve the unpredictable. Indeed Einstein resisted the apparent reality of the discovery of the unpredictable in quantum physics. His statement that "God does not play with dice" is very well known.

3. *The unpredictability of chaos*

While relativity introduces a fundamental change in ideas about space and time without considering the unpredictable, I shall now describe phenomena which involve the unpredictable, without radically changing the ideas about space and time. In fact, the unpredictable is even hidden in Newton's laws! In many situations, which can be very simple, motion is extremely sensitive to minute unmeasurable perturbations. Thus in practice prediction becomes impossible, even when the future is determined by Newton's laws. Such situations are now called "chaos", following James Yorke, who wrote a paper on it in 1975 with Tien-Yien Lee, a student. A popular readable account of chaos is given in the book "Chaos: making a new science" by the scientific journalist James Gleick (Cardinal, Sphere books 1988). It is because of the phenomenon of chaos that the idea of a completely predictable universe based on Newton's laws, which was stated very clearly by the French mathematician Pierre Simon de Laplace (1749–1827), is now known to have been quite wrong.

This form of the unpredictable is already apparent in the calculations made by the great French mathematician Henri Poincarè (1854–1906) shortly before 1900 based on classical mechanics. Not much notice was taken of it until the 1961 computer calculations of the American meteorologist Edward Lorenz, using an extremely simple mathematical model, first showed that even a highly unrealistic model of the weather cannot lead to predictions a very long time in advance. First results were published in 1963. He found that unpredictability could even occur when motion was described by only 3 equations. Other researchers, who did not always know about the work of others in the field, made independent contributions. For instance F. Ruelle and F. Takens came to a similar conclusion to that of Lorenz from the study of turbulence in a fluid in an article published in 1971. Chaos was also found to be important in the mathematical description of the variation with time of the number of living organisms of a particular kind belonging to a certain population. Chaotic processes in different situations, such as those of certain chemical reactions, were described in a book written in French by Ilya Prigogine and Isabelle Stengers, "La Nouvelle Alliance" (Gallimard 1979), without the authors apparently being aware of much of the work already done by other people on chaos theory. It may be noted that previous work done in the Soviet Union is also relevant to this theory.

It is also interesting to note that two of the fathers of chaos theory, Mitchell Feigenbaum and Albert Libchaber, were interested in the scientific ideas of

Goethe, mentioned in the previous chapter. This is perhaps indicative of the radical nature of the change in ideas brought about by this theory.

In order to understand a little what is involved, let us look at the paths of particles, obeying Newton's laws, on which different forces act. A very useful concept in this respect is that of "phase space". This is a completely abstract space, which in the framework of these laws has the 3 dimensions of ordinary space and three dimensions of momentum (mass times velocity in each of the 3 directions) for each independently moving particle belonging to a system of particles which is studied. This means that each particle has 6 dimensions in phase space. So if there are n particles in the system, phase space will have 6n dimensions. It should be noted that if all particles are not independent and/or they are not free to move in all directions, all 6n dimensions need not be taken into account. Now there is a mathematical theorem for particles moving in phase space called the Liouville theorem, which leads to the conclusion that if one studies different possible systems of the same number n particles, with each having a different set of trajectories, the volume of phase space occupied by the different systems does not change with time. What must be emphasized is that the shape of a constant volume is not fixed, it could be the 6n dimensional equivalent of a sphere or, on the contrary, a very long and narrow spaghetti-like region in 6n dimensions, having the same volume. When a chaotic situation occurs, a group of systems of particles at first occupying a small sphere-like volume in phase space and so having both positions and speeds which are almost the same, will later occupy such an extremely thin spaghetti-like volume. The spaghetti-like volume becomes almost "infinitely" thin and is folded a very large "almost infinite" number of times. In this way, corresponding particles belonging to two such systems can become widely separated in space and also have quite different speeds. The volume in phase space which the systems tend to occupy as time passes, will have a "strange" structure, which is called a "strange attractor".

There is a special geometry which describes such a situation associated with the presence of chaos. It is the geometry of fractals, the expression "fractal" having been invented by the mathematician Benoit Mandelbrot. A fractal object will have structure on all scales down to the infinitely small, the same structure being repeated on different scales. The construction of fig. 3.1 shows a very simple example of how to make a fractal figure in the two dimensions of the surface of the page. We start with an equilateral triangle, each of whose sides are divided into three equal parts. Then two sides of a new equilateral triangle pointing outwards are constructed on the central third of each line of the original triangle, this central part being removed. Each side of the figure then obtained is again divided into three equal parts; a new equilateral triangle being erected on the central part, which is then removed. The process is repeated an infinitely large number of times for each side of the figure. In this way an infinitely long figure with structures on infinitely small scales is obtained, but which still fits onto the page! By this process we obtain a "fractal snowflake".

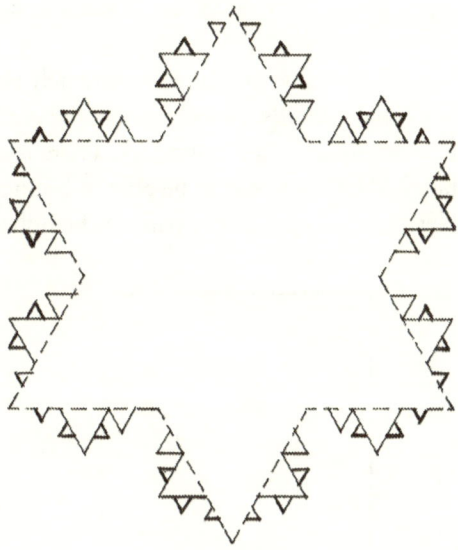

Figure 3.1
The Fractal snowflake after 4 stages of construction. The outer full line is the result of these 4 stages, while the dashed line is the remnant of a previous stage.

Another basic concept connected with chaos is the time scale over which a system becomes chaotic, called the Lyapunov time. The distance between two particle paths which start to diverge from each other in phase space is given by what is called an exponential law. After a short time t, the distance between the paths for two particles which were originally very close becomes equal to $Ae^{(t/\lambda)}$, where A is a constant, e is a very famous number near 2.718 and λ is the Lyapunov time. For times which are much longer than the Lyapunov time, the two paths are very widely separated and the behavior of the system becomes most sensitive to its initial conditions and therefore unpredictable. Thus the system can be said to have lost its "memory" of what were the initial conditions.

The sensitivity of chaotic systems to their initial conditions and to small perturbations can be dramatic and is sometimes called the "butterfly effect". The idea is that the flapping of the wing of a butterfly can completely change the weather at a later date; a distant storm could be produced. The idea is not quite correct, however, as it is impossible to relate the flapping of a particular wing to the production of any particular effect such as a storm. All small and/or distant perturbations have large effects; there is, in fact, a kind of abyss between causes and effects.

It must be emphasized that chaos can occur in extremely simple situations. Fig 3.2 shows what is called Sinai billiards. In the figure there is a billiard table with a flat frictionless surface and a square outer boundary. There is a central obstacle on the table. A billiard ball is then supposed to be able to bounce between the sides and the obstacle. Under these conditions the path of the billiard ball is chaotic.

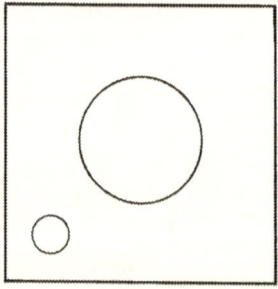

Figure 3.2
Sinai billiards. The billiard ball moving on a flat surface bounces against
the sides and a central obstacle.

Chaotic behavior can be shown in another situation by a double pendulum without friction (fig 3.3). A second pendulum is suspended from a first one. Four types of motion can occur: stationary; periodic, when the ratio of the periods of oscillation of each pendulum is a ratio of two whole numbers; "quasi periodic", when the ratio of the periods is an irrational number, that is, a number which is not equal to any fraction. Examples of irrational numbers are the square roots of 2, 3 and 5 and the number π. And, finally, the motion can be chaotic.

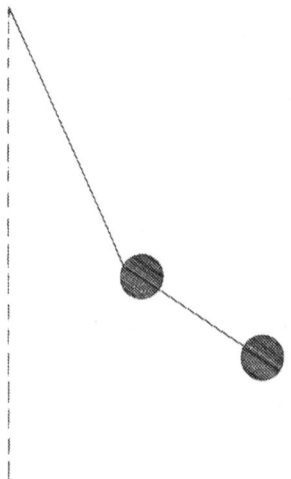

Figure 3.3
The double pendulum without friction, which is also chaotic.

Astronomical phenomena were thought at one time to be particularly good examples of predictable situations. However, chaos is even present in the motions of the bodies of the solar system. The most striking example is that of one of Saturn's satellites, Hyperion. Its shape is far from that of a sphere, its length being about twice its width. The axis and speed of rotation of Hyperion vary in an unpredictable way over a time scale of a few orbital periods (one period is 21 days!). Halley's comet has been fairly periodic for two millennia; however, observations of the comet before the second century BC would be difficult to explain from calculations based on what is known about its present orbit. In fact, the motion of Halley's comet is partly chaotic. Phenomena of chaos also have a strong influence on the properties of the orbits of certain asteroids—the asteroids being very small planets whose orbits mainly lie between those of Mars and Jupiter.

Chaos is also present in the motions of larger planets over very long time scales. The prediction of the motion of Pluto has been shown to be impossible for times of more than about 400 million years. Similarly, predictions of the motions of the inner planets (Mercury, Venus, Earth and Mars) will be completely different than their real behavior more than 100 million years later. According to calculations by Jacques Laskar published in 1994, chaotic effects over a time scale of less than three thousand five hundred million years could even lead to the ejection of Mercury from the solar system, following a close approach to Venus! We might wonder whether ejection of other small planets such as one inside the orbit of Mercury have already occurred. These studies, however, indicate that the orbits of the most massive planets—Jupiter, Saturn, Uranus and Neptune—are virtually unaffected by chaotic phenomena.

There are indications that chaos plays a fundamental role in living organisms, though much remains to be elucidated in this field. As already mentioned, the populations of living creatures of a certain type can vary with time in a chaotic way. A. L. Goldberger And B. J. West suggested in 1987 that the normal functioning of the heart is chaotic—in which case an excessively regular behavior of a person's heart would be extremely serious for his or her health. This idea was expounded in a more popular way by A. L. Goldberger, D.R. Rigney and B.J. West in an article "Chaos and Fractals in Human Physiology" (Scientific American, vol. 262, Nr. 2, p. 34, 1990). In the same way the functioning of the nervous system in general and the brain in particular show signs of the presence of chaos, which may be essential for what they normally do, such as reacting to sense impressions (see "The Physiology of Perception" by W.J. Freeman in the Scientific American, vol. 264, Nr. 2, p.34 1991). Let us note, however, that it is not easy to prove in a mathematically rigorous way that chaos is present. On the other hand, it is at least reasonable to think that many processes inside the body of a living organism satisfy the mathematical conditions for chaos.

Another way of looking at life would be to consider it to be the frontier between predictability and chaos. S.A. Kauffman in "Antichaos and Adaptation" (Scientific American, vol. 265, Nr.2, p. 64 1991) describes mathematical models called Boolean networks, considered by the author to represent the action of genes in living organisms. These models are used to describe how different sorts of cell in the body of a living organism first appear and how evolution proceeds according to Darwin's natural selection. The author mentions different behaviors of such a model, including even a sort of chaotic behavior, when the future development of a model is extremely sensitive to its exact state at a particular time. The author refers to what for many may seem a surprising comparison with the states of matter, suggested by the computer scientist Christopher Langton. According to him predictability is like the solid state and chaos like the gaseous state. The processes of life in such a framework are similar to the intermediate liquid state. This is surprising when it is remembered that in a number of esoteric traditions living organisms are considered to have what are called "etheric bodies", while Rudolf Steiner relates the nature of the etheric to that of water, which is the liquid most commonly encountered by human beings in ordinary daily life. However, it should be noted that the validity of Langton's own calculations has been challenged. We can see in any case that even though the application of the models mentioned may appear to be extremely schematic and materialistic, striking conclusions can be drawn from them, which contain very interesting lessons for us. In particular the idea of life as being between the predictable and unpredictable agrees with normal human intuition; heredity for example, like many other aspects of living organisms, is thought of as being at least partly predictable.

We can conclude from this discussion of chaos that physical systems obeying mathematical laws can become extremely sensitive to infinitesimal, almost "nonphysical" perturbations from the outside. Such systems are in a way "vessels" able to receive what cannot be grasped in the framework of physical predictability, that is, effects which one might think of as being "outside physics". It must be emphasized that as a result of the already mentioned abyss between causes and effects, it is not possible for anybody to physically act on such systems to produce a desired result, unless the systems are continually perturbed over time scales of the order of not much more than the Lyapunov time. This can be understood as being due to the unpredictable effects over longer time scales of any single perturbation at a given time. Living organisms may be good examples of chaotic systems which persist over a long time or perhaps of systems on the border between chaos and predictability. In this way the old arguments about whether or not living organisms possess something not possessed by the non-living, or obey different laws, or, in the language of esoteric traditions, the nature of the "etheric" may in principle be resolvable. Various conceivable implications of such results will be discussed later in this book.

4. *The discovery of the quantum and subatomic worlds*

We now come to that part of twentieth century physics which involves the unpredictable, as well as a major rethinking about the nature of space and time. This physics sprang from the study of various phenomena, mainly the interaction of light with matter, what happens when electricity crosses a near vacuum, radioactivity and, in general, from studies of what occurs over very short distances on very small sub-microscopic scales, as well as what occurs when very small differences of energy are involved. The small distances and small differences of energy cannot be directly perceived by a human being, but only by the measuring instruments. One possible definition of what such small distance scales are at which phenomena occur and which appear strange compared with what humans normally perceive, was once given by R. Burlotte at a meeting in Chatou near Paris. He considered this to be the scale at which the chemical properties of substances no longer existed, that is, scales smaller than that of the molecules from which chemical compounds are made. Let us recall that, as mentioned in chapter two, a molecule of a compound is defined as consisting of a number of atoms. Other definitions of these small scales are, however, possible. For instance, bodies which exist at these scales can keep an electric charge indefinitely, unlike in the everyday world where bodies with a positive electric charge tend to meet bodies carrying a negative charge after a relatively short time. The final result in the everyday world is that both the negative and positive charges are cancelled.

We have already discussed spectrums in chapter two. Now, in order to understand the quantum world discussed in this section, we must look more closely at what kinds of spectrums exist. A very important discovery of nineteenth century physics was that different substances do not emit and absorb light in the same way. Gases, when heated or made to emit radiation by another process in the laboratory, will usually only emit light having certain colours when the radiation is visible to the human eye. In general, if we describe what happens in the framework of the theory of electromagnetic waves, not only considering visible light but also other sorts of radiation, most radiation emitted will have certain well defined wavelengths. Substances which emit at a given wavelength will also absorb radiation coming from the outside at the same wavelength. Dense substances, unlike gases, will emit almost all their radiation over a very wide range of wavelengths. Instruments containing prisms or what are called "gratings", which act similarly to prisms, can be constructed in order to examine the spectra of each substance. By using such instruments astronomers are able to detect the presence of chemical elements in stars and make deductions about the physical conditions of the layers of the star from which the emitted light comes. This is possible because different elements having different conditions will not emit and absorb light in the same way, so the spectra will not be the same. Most of my official scientific work in astrophysics is concerned with the study and interpretation of the spectra of certain rather special types of stars.

The first form of what was to become "quantum theory" originated in 1900 from work of the German physicist Max Planck on the amount of electromagnetic radiation emitted at different wavelengths by a body which is also able to absorb all such radiation falling on it, that is, radiation emitted by what physicists call a "black body". Such a body emits its radiation over a very wide range of wavelengths. Planck found that it was necessary to suppose that this radiation was emitted in separate packets, each containing a finite amount of energy. A packet or "quantum" of radiation emitted at a shorter wavelength than another packet, will contain more energy than that emitted at a larger wavelength. This energy is mathematically expressed as being equal to hc/λ, where h is a basic physical constant called Planck's constant, c is the speed of light and λ is the wavelength the radiation would have if it were crossing a "vacuum", that is, a region containing no matter. Planck's constant is very small, so the separate packets are not seen at the human scale, at which radiation appears to be continuous.

Another phenomenon, discovered by Einstein in 1905, indicates that electromagnetic radiation behaves in certain situations as if it were made of particles. When what is called the "photoelectric effect" occurs, light falling on certain metals will produce electricity. Whether or not electricity is produced does not depend on the intensity of the light, but only on the energy which is contained in each quantum of light falling on the metal. For less than a certain amount of energy no electricity is produced.

In fact, the electricity produced by the photoelectric effect appeared to be carried by particles with a negative electrical charge, called "electrons", the existence of which had been previously indicated by experimental studies at the end of the nineteenth century. These experiments involved the passage of electricity in tubes from which all or almost all gases had been removed. It was by using such methods that certain properties of a single electron were determined by J.J. Thomson in 1897. The electrons seemed to have been separated from atoms which, having lost negatively charged electrons, acquired a positive electric charge. It was with the help of such ideas that the photoelectric effect was understandable: if light could behave as if it consisted of particles which, when they had more than a certain amount of energy before striking the surface of a metal, caused the ejection of other particles, that is, of electrons.

Another phenomenon was discovered by H. Becquerel in 1896. He found that compounds of the element uranium emitted something which could affect photographic plates, even when the photographic plate was separated from the uranium by black paper. Other elements were then discovered to have this property of "radioactivity". It was then found that what was emitted by radioactive elements included electrons moving at very high speeds, electromagnetic radiation having very short wavelengths, (that is, possessing quanta with very high energy) as well as atoms of the element helium with a positive charge. In this way the rules of nineteenth century chemistry were violated by radioactivity. Atoms of a radioactive element produced high velocity atoms of another element, helium, while they themselves were transformed into atoms of a third element. The energy per atom involved in these and other radioactive transformations was, however, much higher than in those of chemistry, in which different atoms combined to form molecules.

The question then arose as to what was the structure of an atom. It seemed to contain electrons and something possessing a positive electrical charge, so that a normal atom is electrically neutral. J.J. Thomson thought that the electrons of an atom were surrounded by the positive charge. This concept was disproved by Rutherford, using the helium atoms with a positive electrical charge emitted by radioactive substances. Two electrically charged bodies having the same charge repel each other, so the positively charged helium atoms coming close to other atoms should be repelled. Rutherford showed that only a small proportion were repelled, some being repelled very strongly. He explained this as being due to the fact that the positive charge of an atom was concentrated in a nucleus, which was much smaller than an atom. The nucleus moreover contained almost all the mass of the atom. It is in the framework of Rutherford's model that the atom and therefore matter began to appear as being almost empty; that is, matter began to appear as something not quite material! Radioactivity came to be understood as due to changes in the nucleus, which was transformed from that of one element into that of another, by emitting positively charged helium atoms, or by emitting electrons, or by sometimes changing in another way.

The nucleus, which could be transformed if it were radioactive or if it were struck by what appeared to be a particle, was considered to contain several particles. One type of particle in the nucleus having a positive electric charge was called a proton, while the other type, which was electrically neutral, was called a neutron. It was first thought that there were electrons in the nucleus, but this idea was abandoned following the discovery of the neutron. As there were only positively charged particles (which, having the same electric charge, should repel each other) and electrically neutral particles in the nucleus, a new strong "nuclear" force was needed to hold the particles in the nucleus together.

Once a "picture" of the structure of an atom had been established, it was possible to study its physics and to further elaborate the quantum theory. According to this picture the electrons of an atom revolved around the much more massive nucleus in the same way that the planets revolve around the Sun. In the case of the atom the forces were electrical, while those attracting a planet to the Sun were gravitational. However, according to classical theory, a negative charge revolving around a positive charge should emit electromagnetic radiation and fall towards the positive charge. An atom with this type of structure should then not be stable. A major step to overcome this problem and also to explain the spectrum emitted by the chemical element hydrogen was taken by Niels Bohr in 1913. Electrons were only able to revolve in well defined orbits around the nucleus. Quanta of electromagnetic radiation were sometimes emitted when electrons jumped from one orbit to another. According to this conception electrons could either spontaneously jump or be made to jump. It was possible to explain many things and not only certain properties of spectra by this "old quantum theory". However, its basis was inconsistent, containing both classical physics and quantum theory. Moreover, it could not explain all spectra. Although the idea of an atom being like a small solar system was soon found to be wrong, it has remained in the popular imagination. It is a very good example of how the thinking of people can be influenced and perhaps even manipulated by a scientific idea which is out of date!

More radical changes in the basis of physics were needed. A most important step was taken by De Broglie in 1923 when he showed that not only did electromagnetic radiation behave as though it were composed of particles, but also that the particles of which matter appeared to be made also had wavelike properties. The wavelength of such a particle equals h/(mv), where h is Planck's constant, m is the mass which must be corrected at high velocities for the increase in its value predicted by special relativity, and v the velocity of the particle (the multiple of m and v is equal to what is called "momentum" in classical physics). This theoretical prediction was experimentally confirmed; these particles could really behave as waves. The mathematical work of Erwin Schrödinger established what is called "wave mechanics", which was able to explain the validity of the previously apparently arbitrary assumptions of the old quantum theory. An atom can only have certain well defined stable wave structures which endure in the framework of this theory and which correspond to the idea of the existence of arbitrary stable orbits in Bohr's theory. Thus an extremely surprising, apparently paradoxical situation had arisen, with two contradictory pictures being needed; one described the world in terms of particles, while the other described the world in terms of waves. The possible meaning of such a result and further developments will be examined in the next two sections.

5. *The meaning of quantum theory*

The paradoxical nature of quantum physics becomes clear when we consider what happens during certain experiments performed to test the various ideas of this physics. Different, apparently contradictory conclusions about what has happened in the laboratory will be obtained, depending on which experiment is performed. For example, certain experiments will show the presence of waves while others will show the presence of particles. It is as if Nature resists the experimenter and gives contradictory answers, depending on which question is asked!

In order to understand the nature of waves in physics, let's look at a very simple situation, shown in fig.3.4. Circular waves on the surface of a lake coming from a disturbance at a point in the lake can only pass through two holes in a barrier. They then reach a side of the lake. At certain points on that side the effects of the waves which have come through each of the holes on the surface of the lake will increase; when waves which have come through one hole tend to raise the surface of the lake, waves which have come through the other hole will also have the same effect. At other times, when waves which have come through one of the holes tend to lower the surface of the water at such a point, waves which have come through the other hole will also tend to lower the surface. Therefore the effects of the waves coming through both holes will reinforce each other at these points on the side of the lake. At other points on the side of the lake the effect of waves which have come through one hole will cancel the effect of those which have come through the other; when waves from one hole tend to raise the surface of the lake, waves from the other hole tend to lower it. At such points the waves may have no effect on the surface of the lake, which then does not move. This is the situation when what is called "interference" occurs, that is, interference between the waves coming through each of the two holes. This can be shown to happen for light and also for the "particles" of modern physics, when the holes are very small. If they only behaved as one expects particles to behave, the situation would be quite different. Let us suppose that small pieces of wood floating on the surface of the lake, which behave like particles, also go through both holes; the number of pieces of wood arriving at any point on the side of the lake will then be the sum of the numbers of those which have gone through each hole. Therefore, waves cannot be particles and particles cannot be waves in classical physics.

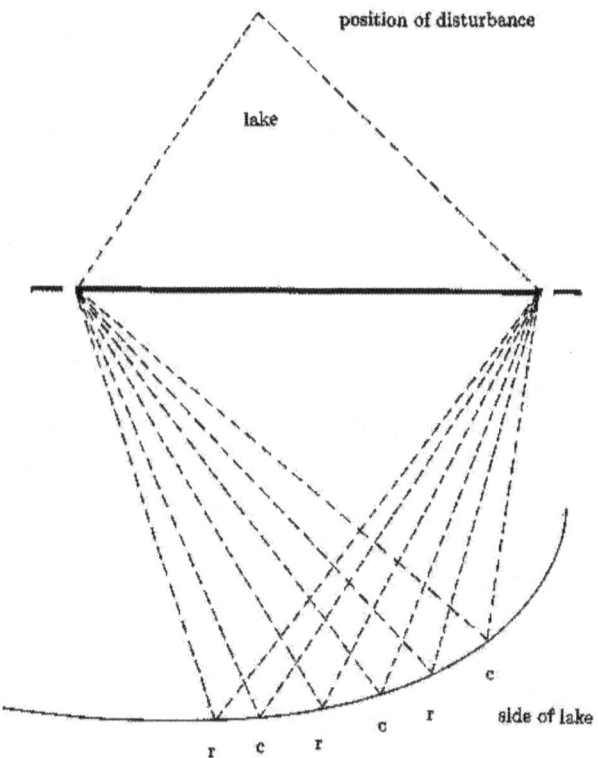

Fig. 3.4—Waves due to a disturbance at a position in a lake which inter-
fere after passing through 2 holes of an obstacle. At the points labeled r
on the side of the lake the waves reinforce each other; at points labeled c
they cancel each other out. The dashed lines indicate the paths taken by
the waves to each point on the side of the lake.

Another characteristic of waves is that a wave of arbitrary shape can be considered to be a sum of regular sine waves, having different wavelengths (for each sine
wave there is a property, which at a certain time varies with position in the same
way that the sine function of elementary trigonometry varies with angle). In this
way, by adding waves with extremely different wavelengths, it is possible to have
what is called a "wave packet", which is very strong in one place and weak else-
where, behaving in some ways like a particle. Adding waves of very similar wavelengths will on the other hand produce a sum which is very extended in space. It
might be thought that one could explain the observed phenomena without
assuming the presence of particles, by supposing that only sums of waves with

different wavelengths are present. This does not work however, as the wave packets would increase in size with time, while experiments show that particle-like behavior can still occur.

As we have seen, quantum physics appears to contradict classical physics. There are objects which can behave both as particles and waves. This contradiction was overcome by Max Born's idea that the waves are not simply waves occuring in matter like those of the surface of the lake, but waves related to the probability of finding a particle with a certain velocity at a certain point at a certain time. This probability can be calculated from what is called the "amplitude" of the wave. In respect to the waves on the lake just mentioned, this amplitude would be the height of those waves. However, it is impossible to know exactly where a particle will be at a certain time or exactly what its velocity will be.

This interpretation in terms of probability can be understood if we consider a fundamental discovery by W. Heisenberg, which will be extremely important for what is discussed later in this book. This is the so-called "indeterminacy" or "uncertainty principle". According to this principle, it is impossible to measure (and, according to the interpretation accepted by most physicists, even to define) with more than a certain accuracy both the position and speed of a moving particle at a certain time. If the uncertainty in the determination of the position x is Δx and the uncertainty in the determination of mass m times velocity v is Δmv, one finds that

$$\Delta x \text{ times } \Delta mv \geq h/(2\pi)$$

In this mathematical expression $\geq$ means greater than or equal to; the Heiseneberg indeterminacy principle states that the left-hand side of the expression can never be less than the right-hand-side. This condition can also be stated in terms of time and energy. If an event requiring an energy E occurs at a time t and Δt is the uncertainty in the determination of t, while ΔE is the corresponding uncertainty in the determination of E: Δt times ΔE $\geq$ h/(2π)

In the case of a moving particle with a velocity v, Δt is in the second form of the Heisenberg indeterminacy principle, the time needed for it to cross the distance Δx, while ΔE is the uncertainty in the determination of the amount of energy contained in its movement (kinetic energy). The second form of the Heisenberg uncertainty principle is also true when an atom emits a quantum of radiation; if the quantum can be emitted at a time within a time interval of Δt, it is impossible to specify exactly what its energy is to an accuracy of more than the value of ΔE.

The resistance to an experimenter by nature mentioned at the beginning of this section can be stated in terms of the Heisenberg uncertainty principle. Nature resists the simultaneous measurement of position and velocity or of time and energy to more than a certain accuracy. The accuracy of each of the quantities which can be measured in a particular situation depends on the type of experiment performed. For instance, some experiments will measure the position of a particle more accurately (if it is known that the particle must pass through a small hole), while others will measure more accurately the speed. As it is impossible to simultaneously measure to an infinite accuracy all the properties of a particle in quantum physics, its future behavior cannot be completely predicted from Newton's laws of motion, which are described in chapter 2 of this book. Only probabilities can be given for what will happen in the future.

In fact, a more detailed examination of the situation raises a considerable number of other problems. Quantum theory is needed to predict what can be observed in a laboratory, that is, it describes interactions between the particles it studies, which have an effect on what happens in the world directly perceived by human beings. The relation between the world of quantum physics and that which is directly perceivable is, however, not easy to understand. On the quantum scale all possibilities of the future behavior following interactions in a system really exist; only one possibility with its associated probability of occurring will be seen in the laboratory. This is clearly seen in the famous example of "Schrödinger's cat". A cat is placed in a box in which there is a capsule containing a highly poisonous gas like hydrogen cyanide. A small hammer can break the capsule when an atom of a radioactive substance in or near the box "decays", that is, it is transformed into the atom of another chemical element, therefore emitting a particle. The particle emitted can be detected by a Geiger counter, which automatically makes the hammer strike the capsule, which then breaks. As a result of this, the cat will die almost immediately. The decay of a radioactive atom is a quantum phenomenon, and the time of it cannot be exactly predicted. At any given time, both the radioactive atom before decay and the decayed atom should exist simultaneously according to quantum theory; therefore that the cat is both living and dead simultaneously should also exist! It is because all except one of the possibilities of quantum theory appear to be eliminated in the human world, that one speaks of the "collapse" of the "wave function" (the wave function being the mathematical description of the waves). The necessity of assuming such an occurrence, which at least at first sight appears to be completely outside quantum theory, puzzled physicists for a long time. In some ways it appeared to be quite illogical.

Einstein could not accept what quantum theory had become, including its interpretation in terms of probabilities, so in spite of the fact that he was one of the "fathers" of this theory, he became opposed to its further development. He thought that real "hidden variables" were behind the probabilities and that they were the cause of the phenomena of quantum mechanics. A consequence of Einstein's opposition was that physicists were told in a scientific paper by Einstein, Podalsky and Rosen in 1935 about what appeared to be a further paradox related to wave function collapse. The paradox is now discussed in terms of what happens when "spin" is measured, that is, something corresponding to the rotation of a particle. More precisely, the spin is called the "angular momentum" of the particle. In classical physics the measured value of angular momentum depends on the direction with respect to which it is measured. In quantum theory the value of an electron, for example, can only be plus or minus half a certain physical constant in whatever direction it is measured. If two particles moving away from each other are produced with spins in opposite directions, having a total spin of zero, the measurement of the spin of one particle in a certain direction will, according to quantum theory, have an immediate effect on what spin can be measured for the other in another direction. Such a prediction contradicts what is expected from classical physics. The two particles can be far apart when the measurement is made; this measurement would appear to have an immediate effect in another place. In this way there is at least a contradiction of the spirit of special relativity, though not with special relativity itself, which does not allow bodies to be accelerated beyond the speed of light. Indeed the basic idea, in which people believe implicitly, that the events of physics are localized in separate places in space and cannot be in two places at once, seems to be in disagreement with quantum theory. Thus quantum theory violates what is called the Bell inequality, which is valid for objects of the everyday world. There is however no mathematical contradiction and the reality of this sort of surprising effect was experimentally confirmed by the work of Alain Aspect, using what are called "photons", that is, the particles associated with electromagnetic radiation, instead of charged particles.

It would therefore seem, at least at first sight, that not only has the unpredictable entered physics, but that in addition the act of measurement itself by a physicist has an effect on a real physical situation. Moreover, the normal laws of space and time of the everyday world inhabited by humans are also violated. It has become very difficult to think that the phenomena of quantum physics could be due to hidden variables localized in separate regions of space. If such hidden variables exist, as some physicists believe, they would appear to be almost certainly not localized. (We should note though, that the interpretation of the experiments showing violation of the Bell inequality is not completely watertight.) There have been many debates about the philosophical implications of the quantum theory. Bohr's "Copenhagen interpretation" tried to minimize the problems, by saying that the theory only described what happens in different experimental situations, in each of which different observations can be made. It was therefore senseless to speculate about what cannot be observed and measured. The wave and particle descriptions needed to account for the different observations made during various experiments were, according to this interpretation, complementary.

It is frequently claimed (though more often by non-physicists than by physicists, a notable exception among physicists being Eugene Wigner), that it is the conscious human observer who causes the wave function to collapse. In this way the observer, after having been eliminated as much as possible, as described in chapter 2 of this book, is again supposed to play an important role in physics. Though I shall describe in the next chapter how we can understand at least certain aspects of quantum physics by the presence of beings with a certain form of consciousness, it still seems to me rather unlikely that the human experimenter can directly produce wave function collapse. The result of an experiment can be recorded by a photograph, which is looked at much later by a physicist. Similarly, an experiment can be performed by a robot. It does not appear reasonable to say that wave function collapse occurs only when one human being, who happens to be the physicist responsible for making an experiment, is informed about the result of the experiment. What happens to Schrödinger's cat when it is not observed? Does a human being need to look at it, to see whether or not it has died? Alternatively, one might think that the cat can also collapse the wave function, but then what would be the simplest living organism able to do this? Alternatively, it has been suggested that wave function collapse occurs as a result of the agreement of all possible observers, but it is, to say the least, hard to make such an idea precise. One might even suppose that an "ultimate observer" or God produces wave function collapse, who would then be presumably outside the physics studied by almost all physicists. Indeed most would very vigorously oppose such an idea. There have in fact been claims that certain experiments show the influence of the human observer on quantum events. In particular, experiments at Princeton University by R.G. Jahn and his collaborators have been cited. These workers have even claimed an observer effect on a previous event, that is, the decay of a radioactive substance recorded by a computer! Much older (1969) experiments led to the claim that human beings could have prior knowledge of the time of decay of a radioactive atom by a sort of extrasensory perception.

These results have not been, as far as I am aware, confirmed by other experimental groups. Even if true, their interpretation might be more complicated than being due to a form of wave function collapse which is directly produced by those performing such an experiment. Such an interpretation is, as we have seen, hard to conceive.

The world of modern physics has often been compared to the descriptions of the nature of the world given by mystics and in particular by eastern philosophies and religions. Bohr, for example, was well aware of a similarity between his conception of wave particle complementarity and certain aspects of Chinese thought. The parallelism of the approach of physics and of mystics is the thesis of a famous book "The Tao of Physics" by Fritjof Capra (Wildwood House 1975, Fontana 1976). Following the Copenhagen interpretation, he describes how both approaches emphasize the basic unity of the world, which is a unity of the divine or, to use Hindu terminology, of Brahman, who is in and is everything. In the Chinese tradition the unity, understood differently, is that of the Tao, which as the way or process of the universe has a dynamic quality. Mahayana Buddhism emphasizes that all things contain everything else. All is interconnected according to such conceptions; we have seen a form of interconnectedness with separation in space being overcome in our discussion of the Einstein, Podalsky and Rosen paradox. Capra also describes the paradoxes or "koans" of Zen Buddhism as being similar to the paradoxes of physics. However, the question arises as to what extent such similarities are real and to what extent they are only analogies. Moreover, at the end of his book Capra himself states that both physics and eastern mystical teachings are needed and that one cannot replace the other when we wish to describe the world.

Another approach is suggested by the definition of the wavelength of a particle. It is, as we have seen, proportional to 1/(the mass of the particle). The corresponding wavelength for a massive body of pre-quantum physics would be much smaller and it has been supposed that the paradoxical effects of quantum physics disappear when the wavelength is below what is called the "Planck length". (The corresponding minimum mass for a body moving near the speed of light is as large as about 2 hundred thousandths of a gram.) The Planck length is a theoretical length below which the effects of gravitation become more important than quantum effects; for this and for smaller lengths space should be completely different from what is known till now in physics. The wavelength of a whole cat will be far below this limit. However, this is not so for small parts of its body, which are much larger than atoms. In the same way, quantum effects could be expected even for specks of dust. Let us note however, that Penrose in the already quoted book "Shadows of the Mind", suggests a more reasonable boundary at much smaller scales of length between the world of quantum physics and that of the everyday world.

A somewhat irrational, completely materialistic way of understanding quantum theory has become quite popular, especially amongst physicists. This is the "Many Worlds" interpretation proposed by H. Everett in 1957, according to which the wave function does not really collapse when a real irreversible quantum event occurs, but rather the universe is split into several universes, which cannot communicate with each other afterwards. Such an irreversible event is defined as being a "measurement", an event which leaves an indication of having occurred at future times. If the "measurement" is the same as what is called a measurement in normal life, like one performed by physicists in a laboratory experiment, each of the different results of the measurement which are possible will be found in a separate universe. In general, each possibility in the evolution of any physical system is realized in a particular universe. The development of the sum of all these universes is completely predictable, although this "predictability" is of no use to an observer living in one of the universes. The paradox of Einstein, Podalsky and Rosen may also be overcome; there is no "action at a distance", as the properties of quantum systems at different points of space in each universe following a splitting must be consistent with each other. It was the Many Worlds interpretation of quantum physics which was referred to in chapter 2 in connection with the anthropic principle: materialists can escape the consequences of this principle if intelligent beings like humans only exist in a small proportion of the separate universes which are created according to this interpretation. Let us note that human beings have clearly no free will according to such an interpretation; all possible actions of human beings will be made in all the different universes in which humans exist. Indeed almost identical, but not quite identical copies of a reader of this book, indeed of all human beings in this universe, would exist in many different universes! We can also note that the Many Worlds interpretation may have become popular among physicists because something more like pre-quantum physics can be valid in each of the universes. In fact, the Many Worlds interpretation could possibly have been inspired by science fiction stories involving "parallel worlds". It appears to me to be a sort of science fiction way of looking at quantum physics.

An interesting way of understanding the meaning of quantum theory has been proposed by Laurent Nottale, who works at the Meudon Observatory in France. In his theory of "scale relativity" he supposes that space has basically a fractal geometry. (We discussed what fractal geometry is in the section of this chapter on chaos.) Nottale proposes that such a geometry is also necessary to describe what happens in quantum physics. In addition, he applies the idea of relativity to lengths, so that no scale of length is to be preferred to any other, in the same way as is mentioned in the second section of this chapter, that no state of motion is to be preferred to any other in Einstein's special theory of relativity. The Planck length plays the same role as the speed of light in Einstein's theory. Nottale is surprisingly successful in reproducing the results of quantum theory. His theory is still far from being generally accepted, but at least shows how a more dynamic conception of space can help to overcome many problems. What is particularly interesting is that he relates quantum theory to the theory of chaos. Indeed he has also applied Schrödinger's mathematical formulation of "wave mechanics" (mentioned in the last paragraph of section 4 of this chapter) to describe the chaotic motions of the planets. I know of no non-technical presentation of Nottale's theory in English; there is one in French which only appeared on page 34 of the September 1995 issue of "Pour la Science" (the French edition of "Scientific American").

Other interpretations of quantum theory as well as attempts to "improve" on it, have also been given. For instance it has been suggested that the wave function of any particle will spontaneously collapse after a sufficiently very long time. A massive object contains many particles, so at least one will suffer wave function collapse quite often. This should provoke the collapse of the others because of the fact that all the waves connected with the particles interact with each other according to quantum theory.

Wave function collapse is now understood by a large number of physicists as being due to the interaction between a system described by quantum physics as having both particle and wavelike properties and the extremely numerous particles associated with a very large number of waves of the large scale world. The latter clearly includes any laboratory apparatus used to carry out a measurement. In the description of the world on the human scale, we do not consider the quantum behavior of every particle associated with waves of which it is made; all we wish to know about it can be described in much less detail. The rest, which is not needed in the description, is called by physicists the "environment". It is the interaction of the quantum world with this "environment", associated with the large scale world, that is thought to cause wave function collapse. For this reason we do not perceive the quantum behavior of every particle inside Schrödinger's cat; the presence of such an "environment" (the body of the cat) is thought to cause the cat to be seen to be either dead or alive, even though it may be poisoned as a result of a quantum phenomenon. This way of producing wave function collapse is called "decoherence"; the idea comes from the work of several physicists, including that of W. Zurek (Readers with knowledge of physics can read a review by Zurek on page 36 of the October 1991 issue of "Physics Today"). It may be noted in this connection that the presence of chaotic phenomena such as those discussed in section 3 of this chapter can also lead to decoherence. Decoherence may be thought of as a way of making real for an observer in the large scale world only one possible history of the universe which has a probability that it will occur. Each possible history must in addition obey the laws of logic of this large scale world. The existence of the phenomenon of decoherence is now supported by laboratory experiments (see article by S. Haroche on page 36 of the July 1998 issue of "Physics Today").

Though all interpretations of quantum physics have not been mentioned in this section, it might appear, at first sight, that everything can be explained by blind unconscious processes of matter, unless claims concerning experiments which could suggest the action of the human experimenter are confirmed. In fact, physicists view matter, energy, space and time in a very different way than in the nineteenth century. These concepts have become very abstract and mathematical. Physicists are still, however, usually materialists; they continue to base their reasoning on the space and the space-like aspects of time. A somewhat different way of seeing a form of soul behind modern physics can nevertheless be proposed, as we shall see in the next chapter.

6. *Other aspects of the nature of matter according to present-day physics*

Present-day physics says other things about the nature of matter. Before leaving this description of contemporary physics, I shall mention a few of them. They are mainly those needed in discussions later in this book.

Firstly, let us consider why, according to physics, different kinds of atoms keep their various structures and indeed why do they not collapse; that is, why don't the negatively charged electrons attracted by the positively charged nucleus of an atom first come very close to it, before falling in? If that happened, matter as we know it should then also collapse. Physics gives two reasons for the stability of matter. Firstly, the wave nature of particles is invoked; the wave structures of wave mechanics described mathematically by Schrödinger's wave mechanics, do not collapse. This can be explained in an approximate though not quite rigorous way using the Heisenberg indeterminacy principle, mentioned in the last section. If we think about an atom of hydrogen, which is understood to have only one electron, the position of the electron relative to the small nucleus would need to become extremely precise when it fell into the nucleus. Now, by the first form of the Heisenberg principle given in the last section, which relates the maximum accuracy to which its position can be defined to the corresponding maximum accuracy for its velocity, the position could only be very precise if the exact value of the velocity were not well defined. In this way the range of possible velocities of the electron can be large enough in such a situation to enable it to have a large enough velocity to escape from the strong electrical attraction of the nucleus!

There is another reason, called the Pauli exclusion principle, to explain why atoms retain their different structures. Only two electrons can have the same wave structure, for which the "quantum state" will be almost the same. This means that when an atom has many electrons, each pair of electrons will have another wave structure. To be more rigorous and precise, the spin of an electron needs to be taken into account. The spin can, as mentioned previously, only be equal to plus or minus a half of a basic constant of spin. The two electrons having the same wave structure will have opposite spins and the quantum state will be the same except for the spin. Let us recall that only certain stable lasting wave structures are possible according to wave mechanics. In this way the different properties of atoms with differing numbers of electrons are explained. In any case, it is clear that all electrons cannot acquire at the same time the same wave structure as close as possible to the nucleus.

Ideas about which particles are fundamental have changed in the last decades. As mentioned in section 4 of this chapter, the nucleus of an atom is considered to contain different particles called protons and neutrons. These particles are not now thought to be fundamental; each of them is believed to contain three fundamental particles called "quarks". Many other kinds of particles, such as the electrons frequently referred to in this chapter, are believed to be fundamental. Such particles are divided into two categories. Those of the first category called "fermions" (named after the Italian physicist Enrico Fermi) obey the Pauli exclusion principle; no two fermions can be in the same quantum state. "Bosons" of the other category (named after the Indian physicist S.N. Bose) do not obey such a principle. The spins of the two sorts of particles are not the same; bosons have spins which are whole number multiples of the basic constant of spin, while fermions have spins which are whole number multiples of this constant plus one half. Fermions include electrons and quarks. Evidence has been obtained by physicists of the existence of three families of fermions. The already mentioned particles called "photons", associated with electromagnetic radiation, are examples of bosons. Let it be in addition noted that each sort of particle has in this framework what is called an "antiparticle" with "opposite properties"; for example the electron with a negative electrical charge has the positively charged "positron" as antiparticle. When an electron meets a positron, both are annihilated, leading to the production of photons. On the other hand, the photon is its own antiparticle. The fact that many sorts of "fundamental" particles have been discovered appears to be somewhat embarrassing, though it is true that the different types of particles which are believed to exist can be classified in a mathematically simple way. Attempts are being made by theoretical physicists to explain all this complexity. At present physicists consider what is called "superstring" theory, to be very promising. A popular account of these things is given in "The Quark and the Jaguar. Adventures in the Simple and the Complex" by Murray Gell-mann (Little, Brown and Co. 1994).

Physicists now explain the forces of physics by the action of different sorts of fundamental particles. This is quite different from the explanations given in the nineteenth century. Let us start with electricity and magnetism, whose description was, as mentioned in section 2 of chapter 2, unified in the nineteenth century by Maxwell's electromagnetic theory. Light was explained as consisting of electromagnetic waves, visible to the human eye. Light was explained as being due to electricity and magnetism. Modern theory reverses this explanation. Photons, that is the particles associated with light and other sorts of electromagnetic waves, are used to explain electricity and magnetism. This explanation also involves the second version of the Heisenberg indeterminacy principle given in section 5 of this chapter, which permits the spontaneous creation of photons, containing energy, for very short times, before they disappear again. When such a particle only exists for a very short time, the uncertainty in the time during which it can exist is very small, but the uncertainty in its energy is then very large. Such a spontaneously produced particle can therefore have a large undetectable energy before it disappears again. If on the contrary such a particle exists for a very long time, the uncertainty in its energy is very small. Therefore only a particle with a small amount of energy can be spontaneously created in the second case. A particle of this sort, which is created in such a spontaneous way and which is not directly detectable before it disappears, is called "virtual". The present quantum explanation of electricity and magnetism then involves the exchange of virtual photons between other particles, which are electrically charged.

The force which binds quarks inside protons and neutrons is called the "strong interaction". It is an indirect effect of this force which is invoked to explain the already mentioned "nuclear force", which keeps the positively charged protons and the electrically neutral neutrons in the nucleus of an atom. Physicists explain the strong interaction by the existence of several kinds of another type of "fundamental" particle, the "gluon"; virtual gluons are considered to be exchanged between quarks, in the same way that virtual photons are supposed to be exchanged between electric charges. This exchange appears to be more complicated than in the case of electricity and magnetism; while only two sorts of electric charges exist (positive and negative), three sorts of the corresponding "charge" exist for the strong interaction.

Another force which influences radioactivity is called the "weak interaction". This is thought to involve the exchange of still other kinds of fundamental particles. The unification of the theory of the weak interaction with that of electromagnetism is a great triumph of recent theoretical physics. The theory of the strong interaction is moreover also mathematically close to that of electromagnetism and the weak interaction. This suggests the possibility of a grand unification of both.

Gravitation is yet another force, which is in principle explainable in a similar way by the exchange of "gravitons". In fact, it is not easy to integrate gravitation in this scheme, though the already mentioned "superstring theory" would appear to be capable of this.

In these explanations of the fundamental forces of physics, the virtual particles which are exchanged are bosons. In this way we can see a polarity according to present-day physics. Fermions which cannot occupy the same quantum state produce the structure of matter, while bosons which can occupy the same quantum state, produce the forces of physics.

The existence of virtual particles leads to very complicated processes. A virtual particle can in its turn give birth to another virtual particle. For instance, under certain conditions a virtual photon can create a virtual electron and a virtual positron (the opposite process to their annihilation). The virtual electron and virtual positron can in their turn create virtual photons; such a succession of processes has no end. Each sort of detectable particle is, according to these ideas, surrounded by a cloud of virtual particles. This "reproduction" of virtual particles may remind the reader of the proliferation of "life" in the normal world of human experience.

Theoretical physicists would like to formulate a "theory of everything" using the concepts of the last three sections of this chapter. This ultimate dream of materialists is clearly not possible. Even if no reference is made to any sort of spiritual teaching, physics only based on space and the space-like aspects of time cannot explain the second and third worlds of Karl Popper and Roger Penrose, described in chapter 1. However, as we shall see, there are still important lessons to be drawn from twentieth century physics. These include, especially, the fundamental roles played by the Heisenberg indeterminacy and Pauli exclusion principles in what modern physics believes to be the basis of matter.

7. *The unpredictable enters mathematics*

I shall now come to what I once heard described as the most amazing discovery of the twentieth century. The field of Mathematics is usually held up as an example of what is certain, of what can be rigorously proved and where it is possible to banish doubt. This view of mathematics was shaken at about the same time as the collapse of the idea of absolute predictability in physics. The consequences of what happened in mathematics are perhaps even more far-reaching. A popular description of the how the previous development of mathematics led to this "disaster" is given by Morris Kline in "Mathematics: the Loss of Certainty" (Oxford University Press 1980).

In the same way that nineteenth century physicists wished to predict everything from the laws of physics, mathematicians wished to be able to deduce all possible results in mathematics from a few fundamental principles. A good example of how this can be done is elementary geometry, which was developed by the ancient Greeks. Euclid proposed a number of "axioms", that is, fundamental principles on the basis of which all kinds of geometrical results called "theorems" can be proven. For instance, it is possible to prove in such a way the famous Pythagorean theorem of for a right-angled triangle, The square of the length of the hypotenuse equals the sum of the squares of the lengths of the other two sides (fig 3.5). One of Euclid's axioms about where two straight lines, crossed by another line, will intersect has, in fact, an uncertain validity; if it is dropped, other "non-Euclidean" geometries, discovered in the nineteenth century, become possible.

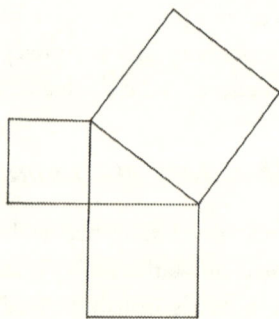

Fig. 3.5—Pythagoras' Theorem. The square of the longest side of a right-angled triangle is equal to the sum of the squares of the other two sides.

By the nineteenth century many concepts had entered mathematics which could not be logically justified; it was then that mathematicians started seeking a logical basis to the whole of mathematics. A fair amount of progress was made at that time. In the early years of the twentieth century, different schools existed with different opinions about the foundations of mathematics.

One school included G. Frege, Bertrand Russel and Alfred North Whitehead. They tried to base mathematics on the laws of logic. Numbers are the basis of mathematics, so in order to possess secure foundations, it was necessary to find a logical basis for the concept of what a number is. However, the attempt to base mathematics on logic led to a number of difficulties; in particular certain axioms of doubtful validity were needed. In the end, Bertrand Russel himself admitted the failure of the attempt.

Another school, opposing the attempt to base mathematics on logic and led by L.E.J. Brouwer, a Dutch mathematics professor, was called "intuitionist". Brouwer asserted that mathematics was an activity of the human mind which had no real existence outside it. The mind is able to have direct intuitions of mathematical principles. Intuition then determined the soundness and acceptability of ideas, which were neither determined by experience nor by logic. Acceptable logic is to be based on mathematical intuition. The idea of deducing mathematics from basic axioms was rejected. The result of this type of approach in the end was, as might be expected, that many mathematical methods and results could not be accepted.

The leader of a third "formalist" school was the German mathematician David Hilbert. He disagreed with the idea of basing the concept of whole numbers on logic; for him whole numbers were already implicitly present in logic from the beginning. He was also alarmed by the intuitonist approach, which rejected large parts of mathematics. Hilbert wished to found mathematics on basic axioms which needed to be consistent with each other. These axioms involved both mathematics and logic. He proposed to use a special logic to show that there were no inconsistencies in mathematics.

In any case, two basic problems concerning the foundations of mathematics remained in 1930. The first was to prove that it was impossible to obtain contradictory results in mathematics, that is, that mathematics was consistent with itself. The other problem was how to establish a complete set of axioms for any branch of mathematics from which everything could be derived. In 1931 the Austrian mathematician Kurt Gödel showed that it is impossible to resolve these problems. It is not possible to prove that there are no contradictions in any mathematical system which includes the arithmetic of whole numbers! In addition, any theory of whole numbers is incomplete; that is, that if one has a finite number of axioms there are mathematical statements which can be neither proved nor disproved! It was in this way that a sort of indeterminacy entered mathematics, as it had entered physics.

In 1936, Alonzo Church showed that in general there is no way of deciding in advance whether a mathematical statement is provable or not. As a result, mathematicians cannot have standard procedures for proving things, even though they may be provable. This can also be understood as stating that they have to work to prove things; they are not in danger of becoming unemployed.

A good example of such problems is Fermat's famous last theorem, stated by the seventeenth century French mathematician Pierre de Fermat. It concerns what happens when three whole numbers are raised to the power n; that is, are multiplied by themselves n-1 times. The question is when do three different whole numbers A, B and C exist, which satisfy the following expression:

$$A^n + B^n = C^n$$

This is clearly possible when n equals 1, so they are not multiplied by themselves; all whole numbers except zero, one and two can be set equal to sums of two other different positive whole numbers. When n equals 2 this is still possible in certain cases, such as 3 times 3 plus 4 times 4 equals 5 times 5, or:

$$3^2 + 4^2 = 5^2$$

another example is:

$$5^2 + 12^2 = 13^2$$

Fermat's last theorem is very simple; it states that when n is greater than 2, no whole numbers exist satisfying such a relationship, unless either A, B or C is zero. Though Fermat claimed to have found a proof, a rigorous proof was in fact only found with difficulty by the Princeton University mathematician Andrew Wiles in 1994. At one time the theorem was suspected of being unprovable.

The results of Gödel and Church indicate that no mechanical procedure can be invented to make all mathematical proofs. A result is that all such proofs cannot be made by a computer following fixed rules. The English mathematician Alan Turing, who was also one of the fathers of the theory of computers, conceived a model of an ideal computer, which is called a Turing machine. Such a machine can be considered to have solved a mathematical problem if, after starting the calculations needed to solve it, it comes to the end and stops. This type of machine can, in many situations, continue to compute without ever stopping, without it being possible to give a general rule about when this will occur. In fact if a general rule existed, a basic contradiction would occur in certain cases; it would be possible to prove for certain mathematical statements that it is impossible to prove or decide whether a particular statement is true or not! As a result the general rule cannot exist, as is explained by Penrose in "Shadows of the Mind". Such reasoning is similar to that used to prove Gödel's theorem itself. This type of difficulty arises because a mathematical statement can be about itself, that is, a process of thinking can be about itself.

Arguments of this type are used by Penrose in "The Emperor's New Mind" and "Shadows of the Mind", to show that computers, constructed according to known principles, cannot reproduce all human thinking. Such a computer can only follow well defined rules. Penrose supposes, however, that a computer working according to the rules of quantum physics might overcome this problem. He suggests that the human brain functions as a machine of this type. In "Shadows of the Mind" he proposes a rather speculative model involving the action of quantum physics in the working of the brain's nerve cells. However, he does not prove that human thinking can be reproduced in such a way.

The first conclusion which can be drawn is that although many aspects of thinking can be imitated by mechanical processes, it is extremely difficult to imitate all aspects of thinking in this way. If the brain is only a predictable mechanical system, it would appear to be incapable of being completely responsible for producing all of what is involved in thinking. We may then suppose that the brain is not able to think about thinking itself. It appears to me that the only possible way one might try to escape from such a conclusion would be to invoke the probably unpredictable chaotic nature of the brain, which as a result of minute perturbations to it could then arrive at thoughts, which do not directly follow from previous thoughts…However, it would appear to most people that thinking about thinking is clearly not an irrational process!

The basic importance of thinking as an autonomous activity was emphasized on a much more fundamental level by Rudolf Steiner, particularly in his "Philosophy of Freedom", mentioned in chapter 1. He pointed out that thinking is a basic human experience and showed that man's knowledge of the world is obtained from both observations of what is perceived and thinking, each being necessary right from the beginning. Thinking is in this framework an activity directed at the perceptions of an object observed by the self of a human being, who is immediately aware of this activity. The content of thoughts and how one passes from one thought to another cannot be governed by any process involving the physics of the brain, but must be only governed by the laws of thought. This is clearly the case in any sort of scientific activity, where it is impossible to escape from thinking. In his Philosophy of Freedom, Rudolf Steiner wrote: "My observation shows me that in linking one thought to another there is nothing to guide me but the content of my thoughts. I am not guided by any material process in my brain. In a less materialistic age than our own this remark would of course be entirely superfluous". According to Rudolf Steiner, it is in the world of pure thinking and in actions inspired by this activity, that a human being can be free.

Gödel's theorem suggests even more striking conclusions about the nature of mathematics and the world of pure ideas studied by it, to be discussed in the next chapter.

4

Where May the Action of Soul be Found?

1. *The nature of the quantum and subatomic worlds*

We have seen in the last two chapters how the development of the sciences in the last few centuries has gradually led scientists more and more into the study of strange worlds very far from that of the everyday world of human experience. Basic assumptions were made starting at the time of Newton, about what really was and what really was not scientific. Physics was based on the study of bodies which exist in space and time, only taking into account their spatial properties and their properties in time as measured by clocks, that is, those properties which can be described by the space-like aspects of time. In addition physical phenomena were studied to a greater and greater extent by instruments, thus bypassing direct human perception as much as possible. Moreover, as the power of instruments was increased, it became more and more possible to study many phenomena which cannot be studied by other methods, such as those which occur on very small scales, which are generally believed by physicists to be fundamental. Therefore, what was studied was the interaction of the matter participating in a phenomenon with the matter associated with a measuring instrument. The following of this path by physics led to the crossing of a threshold at around the beginning of the twentieth century; we described in chapter 3 discoveries made after this threshold was crossed. The laws of other sciences including, for instance, biology, were supposed to be the consequences of the action of the laws of physics in various situations. As a result, the soul world of inner experience, that is the second world of Popper and Penrose, was eliminated, leading, as we saw in chapter 2, to quite paradoxical conclusions about the nature of the world. We shall now look at the discoveries made by the sciences in the twentieth century in different ways than those in which they have usually been looked at up till now, to see whether hitherto unknown soul aspects are not hidden behind them.

We shall in particular look for the presence of the three aspects of the worlds of inner experience, discussed in section 4 of chapter 1. A certain effort of thinking will be required to do this, though it is not necessary to go into the technical details, which are only of interest to experts.

Let us start with the world of the very small, described by quantum physics, which as we shall find can teach us some very basic lessons. It is, as we saw in chapter 3, a world where different possibilities of successive events occurring co-exist; one of the possibilities will be realized in the large scale world, by what is called "wave function collapse" or according to the interpretation physicists now mostly accept, by interaction with the environment. Exactly what will be produced on the large scale by events on the small scales of the quantum world is unpredictable; in order to make the sort of prediction possible using Newton's laws, it would be necessary to know the values of physical quantities, that is, to be able to make physical measurements to an infinite accuracy (see the end of section 1 of chapter 2). This is impossible according to Heisenberg's indeterminacy principle. However it must be emphasized that the physics of the large-scale world is still closely connected with that of the small scale world; only certain possibilities are allowed by that world, which plays a fundamental role in the ideas of physicists about the nature of the processes of the world which can be directly perceived by humans.

As we saw in section 5 of chapter 3, the Heisenberg indeterminacy principle can be stated in different ways. It is a limit to the accuracy to which certain physical quantities can be simultaneously measured and even be thought of as having a meaning in physics. In the second form, the principle can be stated as being a limit to the accuracy in the measurement of the time of an event, multiplied by the accuracy in the simultaneous measurement of energy which must be present when the event occurs. Two examples of what is meant by an event and its energy in this context were given in that section of chapter 3. We spoke there about measurements being resisted by nature. More precisely, if what occurs is examined more closely we find that when an event which happens in a physical system is studied using a measuring instrument, we can suppose that there is resistance between the matter undergoing the event and the matter of the instrument with which it interacts. This can be seen because an experiment using instruments will measure more accurately the time of an event and less accurately the energy, while different experiments using other instruments will measure more accurately the energy and less accurately the time. Similarly the position of a particle will sometimes be measured more accurately and in different experiments the speed will be measured more accurately; there is also a resistance to both the latter quantities being simultaneously measured to an infinite accuracy. It is this basic resistance which limits the accuracy of measurement, while the highest possible accuracy in measuring all the properties possessed by any physical object according to pre-twentieth century physics, would be needed to predict its future behavior and so to know everything about it.

The Heisenberg indeterminacy principle is indicated in other sorts of resistance, described in section 6 of chapter 3. When combined with the Pauli exclusion principle, it can be partly related to a resistance between electrons and the nucleus of an atom, which prevents the electrons from falling into the nucleus and so maintains the atom's existence. In addition, atoms resist each other in their interactions. We saw in the same section of chapter 3 that the Heisenberg indeterminacy principle is also invoked in modern theories of the fundamental forces of nature (which are electricity and magnetism, the strong interaction, the weak interaction and gravitation); the principle permits the creation of "virtual particles", which produce these forces. The existence of the fundamental forces in nature may therefore also be thought of as a kind of resistance acting between particles. We may conclude that the Heisenberg indeterminacy principle is directly connected with the most basic characteristic of matter, which can also be observed on the scale of normal human experience. Matter resists matter; bodies made of matter act on other bodies made of matter in many different ways and so force them to have various sorts of behavior. For instance, to take a rather extreme case, it is very difficult to make a solid body pass through another solid body! It is possible to see in a more precise way how the Heisenberg indeterminacy principle is related to resistance and this is especially useful for its examination in the following discussion. We shall express it differently by making an extremely simple mathematical manipulation, that is, by taking the reciprocal of both sides of the expression in the form relating time and energy, given in the last chapter (this means that we take 1/each side of that expression). The result is:

$$1/\Delta t \text{ times } 1/\Delta E \leq 2\pi/h$$

Here the symbol $\leq$ means less than or equal to, so this new mathematical expression states that the left-hand side must always be less than or equal to the right-hand side, the left-hand side containing the degree of precision of the value of time $1/t$ multiplied by the corresponding degree of precision in energy $1/E$. In the last expression the time is more precise if $1/t$ is large, while the energy is more precise when $1/E$ is large. These precisions are limited by what we have just seen can be thought of as a form of resistance; their multiple can never be greater than the physical constant given on the right-hand side.

We will now try to see what sorts of soul processes and actions of conscious beings can be behind this basic nature of matter as studied in physics. It will be necessary to examine more closely the three aspects of the soul possessed by many sorts of conscious beings, as described in chapter 1. Let us start as a first step by looking at what the simultaneous existence of different beings means. In fact, if more than one conscious being really exists, there must be a certain degree of separation between the different beings. This means that each being must possess something that is unpredicted and uncontrolled by other beings; on the other hand, the various beings can have relationships with each other. Each can then communicate something new to another being. If on the contrary everything in the world were under the control of just one being, a kind of supreme cosmic "dictator" (or God according to the ideas of religious extremists), it could in principle decide everything, as nothing else could have either the ability or even the desire to thwart it!

In order to see more clearly what is the role played by the three aspects or abilities of the soul, let us consider what happens if a being has a desire for something to happen. In order to realize the desire it needs the ability to act. However, it must also know what will be the results of any particular action; otherwise an action need not produce the desired result. This indicates that a desire is not enough; it must be accompanied by the two other soul aspects. Many examples of this exist in the most ordinary events of everyday life. For instance, suppose one wishes or desires to cut a circular cake into two equal halves; one must have the ability to use a knife, but also know enough about the geometry of a circle in order to place the knife in the right position and cut the cake correctly. If on a more difficult level one wants to build a house it is necessary to have available the force required to do the building work; but it is also necessary to know something about mechanics, so the house does not collapse. When a person wishes to make a journey by plane it is not only necessary to be able to go to the airport, but also to know where the airport is…We can conclude that without the ability to act, any amount of knowledge would be useless, while without knowledge any amount of ability to act would also be useless. In any case it is easy to see why one might expect the three abilities of the soul to be present very often in some form or other, when different conscious beings exist.

When independent beings exist, they can be expected to have different desires; it is in such a way that resistance can arise between them. Indeed the experience of resistance due to beings of the outer world might even be thought of as what gives rise to the consciousness of a particular being. Let us suppose that a class of beings exists who, when they interact with any other being of the same nature, always resist it to the same extent by limiting both its ability to act and its knowledge. In this way there could be a limit to the amount of ability to act multiplied by the amount of knowledge of each of these conscious beings and so there could be a limit to the realization of what is desired by each of them. In this way the feeling of contentment or happiness of each being would be limited. In that case we could have a situation where: the amount of knowledge times the ability to act is less than or equal to a certain value.

This is clearly very much like the last expression for the Heisenberg indeterminacy principle. This is easily seen because, as mentioned in section 2 of chapter 2, in physics energy is a form of ability to act. What is now involved is that if to produce a certain physical event one must not have either too much or too little energy, the ability to act will not be proportional to the total amount of energy available, but rather proportional to the precision with which the energy can be applied. Such situations can indeed exist in everyday life; when for instance a person in an office wishes to throw away some sheets of paper, in order that they fall into a wastepaper basket they need to be neither thrown too fast (with too much energy) nor too slow (with too little energy). The precision in the determination of the time of an event in the Heisenberg uncertainty principle can then be considered as being a form of the amount of knowledge which is available.

Our discussion indicates that the resistance of the quantum world, associated with Heisenberg's indeterminacy principle, is understandable as due to the interaction of conscious beings, which resist each other. According to this viewpoint the consciousness of such beings is, unlike what many people believe, not directly related to the consciousness of a human observer. These beings must in such a framework act when interactions occur between the objects which physicists think of as being very small particles. The beings, like these particles, must then not have the same relationships to space and time as the bodies and living organisms of the world of human experience. In this case, the physical constant on the right-hand side of the last mathematical expression for the Heisenberg indeterminacy principle must represent the maximum happiness attainable by each of the beings involved, when it uses all the means at its disposal to attain it. What for a human being would be a feeling of happiness is then limited by a constant of physics as the small scales of the quantum world are approached. We can now form a vivid picture in our minds of what appears to be a heartless inhuman world, which is the one we are considering here. There is no real love, but only a sort of resistance exists between beings, so the happiness of each of them is limited by or "imprisoned" inside a constant of physics.

It appears therefore that the small scale world we are now examining is not only without most of the qualities of human society, but that it is also without the qualities of nature as we directly experience it. Relations between animals like those between human beings (and between human beings and animals) are clearly not governed only by blind resistance. Animals have sexual relations and they often live in groups or herds. The social insects such as the ants and bees live in large communities. Symbiotic relations, where individuals of different species live together and help each other, also exist in nature. In this way the domain where the Heisenberg indeterminacy principle applies would seem to be at a level which is "lower" than the nature of human experience or, to use Rudolf Steiner's expression, a world of "sub-nature". In his last letter to those connected with the "anthroposophical" movement founded by him, written in March 1925 (see "Anthroposophical Leading Thoughts", Rudolf Steiner Press, 1973), he says that technical science and industry have entered a realm of sub-nature, from which among other things electricity is derived. At this point we may recall, as stated in chapter 3, that in the small-scale world we are now considering, electrical charges persist indefinitely.

The world in which the Heisenberg indeterminacy principle acts with what we can understand as being blind resistance and the "imprisonment" of happiness inside a constant of physics, appears to us as a world without love and without morality; it is, however, not immoral but rather amoral. In any case, it cannot be thought of as a "good" world. At this point it is possible to go much further in understanding that world, by using certain conceptions formulated by Rudolf Steiner to explain the phenomena we are considering. He often describes what can become two sources of evil for human beings when they are not faced in the right way. These sources of evil which oppose human beings are, according to him, two beings which have opposite natures and whom he calls, taking names from different religious traditions, Lucifer and Ahriman. According to Rudolf Steiner, Lucifer, which in the Christian tradition is a name for the devil, tempts a human being through pride and through producing all sorts of illusions about the apparent beauty and wonderful nature of many aspects of the world around him or her, which in reality are very different, such as false ideals. Ahriman or Angra Mainyu is the name of the devilish being of the Zoroastrian religion, which was practiced in Persia before the rise of Islam. According to that religion, as a spirit of evil, he resists Spenta Mainyu, who is the son of the supreme God Ahura Mazda (see "Zoroastrianism: the Religion of the Good Life" by Sir Rustom Masani, Collier Books, New York, 1962). Let it be noted that Zoroastrianism is sometimes described somewhat differently; I have here mainly followed the description in the book just mentioned, written by a twentieth century follower of this religion, that is, by an Indian "parsee". Rudolf Steiner gave the name of Ahriman to a being associated with matter, the materialism and philistinism common among human beings, who also tries to imprison human behavior through rigid rules and through what is cold and impersonal in the world. According to Rudolf Steiner both Lucifer and Ahriman are necessary elements in the evolution of the world; a human being must strike a balance between them. I should also point out that Lucifer and Ahriman have only been described here in a very simplified way; Rudolf Steiner's descriptions of them are much more subtle, involving many situations where both are present and complex interactions occur between the two.

In the framework of these two sources of evil the Heisenberg indeterminacy principle has an important role. The world where this principle is important can be clearly pictured as being a world of cold resistance with happiness imprisoned inside a constant of physics, and can also be conceived of as a world of Ahriman. Indeed, it was through such a picture that the author of this book was able to more clearly understand what Rudolf Steiner meant when he discussed the nature of Ahriman.

Quantum physics is closely connected with the physics of the world of human experience and in particular with the human experience of matter. The physics of the very small acts on the physics of the world in which we live. In fact certain spiritual traditions have considered matter to be evil and to be the domain where the devil acts. In this connection we should mention various Gnostic movements which existed in the first few centuries of the Christian era, the Manicheans (from the third century onwards) and the Cathars. All these movements were violently opposed by the Church. Catharism, which became widespread early in the eleventh century, later became particularly influential in south-western France. The Cathars were heretics as far as the Roman Catholic church (and also the Orthodox Church) was concerned, to be brutally exterminated by burning its adherents at the stake. Though great crimes were committed against such movements, we need not agree with all that they stood for; in particular their view of the world appears to be somewhat one-sided. As already mentioned, the nature in which we live is not the same as the world of quantum physics; nature does not seem to contain only evil.

As is well known by people who have some acquaintance with science, the other aspects of nature studied by physicists do not show love or morality either. This amorality of physics leads to a kind of freedom for human beings in the world on the human scale. It is not necessary to "persuade dead objects" to do what we want them to do by being "kind to them"; we can make them do what is possible according to the laws they obey. To give an example, a chair will not play games with somebody who wants to sit on it, as happened in a film made by students I saw many years ago.

Before ending this section, it is useful to briefly look again at the Pauli exclusion principle. As we saw in section 6 of chapter 3, "fermions", which are the particles which obey this principle, produce the structure of matter. No two particles can have exactly the same properties. In the framework of the present discussion concerning the possible soul properties of beings acting in quantum physics, this might be understandable if it is considered as being due to a refusal of any being connected with fermions to imitate another such being. Such a property might appear to be another kind of resistance. We shall now examine some other aspects of science before returning to quantum physics later in this chapter.

2. *Where can soul be found in the phenomena of Chaos?*

As stated in the last section, the nature which is inhabited by human beings does not have the properties of the quantum world, though the latter clearly has what is an extremely important role in determining the physical processes of the world on the human scale. It is possible, however, as we shall now see, to understand certain aspects of nature if we consider it to be the stage on which other kinds of conscious beings can also act. Such beings include humans, who have an inner experience of being free to act, as well as many other types of beings. These beings must possess different forms of what for humans are the three soul abilities.

As we have seen, according to present interpretations of quantum theory, different possibilities of successive events co-exist in the quantum world. All but one are eliminated at the boundary between that world and the world we experience; as a result a unique sequence of events occurs in time, which must obey the logic of our large scale world. The nature of time, already discussed at the end of chapter 2, is in fact essential for understanding that world on the human scale.

We shall now return to the examination of what is called "chaos". In situations of chaos which can occur in the world directly experienced by human beings, physical events are unpredictable because their occurrence is extremely sensitive to the exact physical conditions and indeed to any minute perturbation from outside. As stated in chapter 3, chaotic systems can be thought of as being "vessels" able to receive what cannot be grasped in the framework of physical predictability. For instance, there are indications that chaos is present in living

organisms, which could be such "vessels"; something could be present in the phenomena of life which cannot be grasped in this framework of physical predictability. The future development of such a chaotic system would, among many other
things, be dependent on the exact time at which a particular small event in the
system occurred. In this way a system in a state of chaos, like a system obeying the
laws of quantum physics, also resists attempts to make it have a particular behavior. However, we shall now see that this resistance can be directly connected with
the nature of time. As described in chapter 3, chaos plays a role on time scales
much longer than the Lyapunov time; it is difficult to influence the behavior of a
chaotic system on such a time scale. This type of system is so sensitive to minute
infinitesimal perturbations, that the results of an action very close to a particular
time are unpredictable on such relatively long time scales. Indeed, if these time
scales are sufficiently long, the behavior of the system is influenced by the indeterminacy of the Heisenberg principle.

The question is—Is it possible for conscious beings possessing soul abilities to
act on chaotic systems such as those of living organisms? Such beings would need
to have behavior which could not be completely predicted and controlled by
other beings, that is, which would have to be not predictable according to the
laws of pre-twentieth century physics.

The control of chaos has interested a number of scientists (for a mathematical
treatment see "Using Small Perturbations to Control Chaos" by T. Shinbrot, C.
Grebogi, E. Ott and J. A. Yorke in "Nature" vol. 363, p. 411 1993). It is possible
to apply a succession of small perturbations to make a chaotic system stay in the
same state, in order to overcome the consequences of changes due to small influences from outside the system. Alternatively it is possible to slightly change basic
properties (mathematical constants) of the system, so that in its unpredictable
behavior a state near that which is desired is attained fairly quickly, without using
large quantities of energy. This type of manipulation clearly has practical applications, for instance in changing the chaotic orbits of spacecraft.

A being who wished to influence a chaotic system on a time scale much longer
than the Lyapunov time, without being hindered by the rigid laws of cause and
effect of pre-twentieth century physics, that is, without needing significant quantities of energy produced by physically predictable processes, could not use such
methods to control chaos. It would appear that if such beings existed, they would
need to "see" what the results of any possible action are, that is, to "see" and also
act in the future! In that way they would be able to "select" the future they
wanted. Their perceptions would need not to be confined to one instant of time;
these perceptions would have to be "extended" over an interval of time. In fact,

the satisfaction or feeling of happiness obtainable by such beings would be proportional to their ability to act times the ratio of the amount of time they could "see into" the future divided by the Lyapunov time.

In many situations of this sort the ability to act would no longer be the precision with which an energy can be applied as in our discussion in section 1 of this chapter about how to understand the Heisenberg indeterminacy principle, but rather the amount of energy required to perform the action. If the ratio of the two times were sufficiently large, knowledge of the future would be very great and the energy required would be very small or even in the domain of quantum uncertainty. In such a way, beings of the sort we are now considering would, it appears, be limited in the satisfaction they could obtain by something which again looks like the Heisenberg indeterminacy principle, though knowledge and ability to act are now represented by different physical quantities. The three aspects or abilities of the soul world described in chapter 1 again appear here. However, the limit to the present kind of satisfaction or happiness is not necessarily a constant.

Some readers might at this point think that the author of this book has become completely irrational! Why think of beings able to "see" the future, why think that any other beings act in chaotic systems? Why suggest something as farfetched as beings whose perceptions are extended in time? A simple reason for considering this kind of possibility is what is directly experienced by a human being. A human has the experience of being able to act, of being able to use his or her will to at least partly obtain what he or she wants. In particular, it is possible to act on the body, to move the limbs for example. As described in chapter 3, there are indications that certain processes in the human body and particularly in the brain are chaotic, that is, sensitive to infinitesimal perturbations from outside. The present discussion offers a possible framework for a human to possess something which, in order to act on the body and control it, "sees" very slightly into the future or which is slightly "extended" in time. Other living organisms are also partly chaotic, or are at the border between chaos and predictability, so the same situation might to some extent also exist for them. The same could also be true for some other examples of chaotic systems which are not thought to be associated with life. In this way beings which are not governed by the laws of deterministic physics could be present in the world and possess aspects corresponding to the soul abilities. This is far from a rigorous proof, but rather something which appears to be not completely unreasonable if we free ourselves from conventional materialistic ways of thinking.

We can see two possibilities if we consider the role of chaos in living organisms: either the situation is as suggested here, that the world of human inner experience or the soul is really significant and at least partly independent of "normal" physics, ; or the matter of which the human body is composed plus the matter outside the body which can produce minute perturbations in it, act in their own unpredictable and uncontrollable ways. Matter would, according to the second possibility in the framework of what might be seen to be a more "consistent" sort of materialism, have to possess additional soul abilities, which however are not those which can be thought of as being behind quantum physics. The alternatively even less satisfactory materialistic explanation is that the human experience of being able to act is only an illusion. Though we cannot rigorously "prove" that matter in a state of chaos is acted upon by beings extended in time, what is here suggested does not, in view of such considerations, appear to be irrational.

In fact the present conception of the action of beings extended in time on chaotic systems might also help to resolve an old philosophical problem of what is called "dualism". If both a material body and a non-material "soul" exist, as is the case according to dualism, it is not clear how this soul could act on the body. However, it may be possible to overcome such a difficulty if what is involved is a being extended in time, acting on a chaotic physical system. In the framework of the point of view presented here, strong interactions can be expected between a chaotic brain and what is extended in time. In such a way, quite a good correlation could exist between brain processes and the world of inner experience, which is then "reflected". Studies of the brain do indeed show such a correlation.

At this point certain objections might be raised against what is being proposed here. The three soul abilities appear at first sight to be directly related to consciousness, while modern psychology emphasizes the role of the unconscious and subconscious, at least in humans. However it is easy to realize that the conscious human being need not be alone even in his or her body, whose chaotic processes may be also continually influenced by many other beings. The latter could then be expected to be unpredictable and uncontrollable, being able to resist the human, by limiting his or her amount of knowledge and ability to act. Such beings then can be thought of as influencing the unconscious and subconscious. The effects of such influences would be complex, thus explaining the complexities of human psychology, including mental illnesses. If a human felt guilty about a past action which he or she had performed, for example, it might be possible to "forget" it, while other beings "retained" the knowledge contained in the memory. The awareness of this memory or apparently irrational actions based on it, could then be "returned" to the human in certain circumstances.

There are reasons to believe that the soul abilities of a human being have quite subtle relationships with time. As already stated at the end of chapter 2, willing is connected with action on the future from the present, while it is possible for the human in the present to reflect on the past. In the present the possibility to act on the future dies. These considerations might suggest that things are rather more involved than stated up to now, if what is happening is the action on chaotic situations of a being extended in time. The human action of willing could in this case act on times which tend to be slightly to the future of what is experienced as the present, while the person would only be able to consciously see and think about what tends to be slightly in the past. See figure 4.1

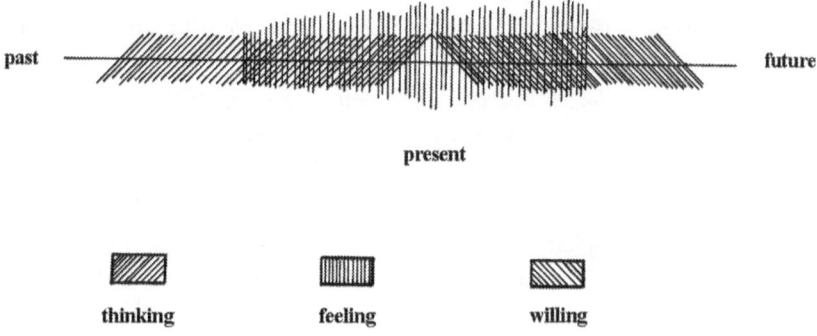

Figure 4.1—Suggested separation of thinking, feeling and willing of a human being in time

This suggestion is not, as it might first appear to be, empty speculation, as it may be possible to explain in such a way the surprising results of certain psychological experiments involving stimulation of the brain, mentioned in particular in the book "Shadows of the Mind" by Roger Penrose. Among the people who carried out these experiments, we should mention H.H. Kornhuber, while later ones were performed by B. Libet. Kornhuber's experiments indicated, among other things, that if a human being decides to perform an action, the electrical activity of the brain starts to change one to one-and-a-half seconds before he or she is conscious of having made the decision to act. It may be possible to understand this result from the nature of that which in a human being is, as suggested here, slightly extended in time, with a will acting slightly in the future on the brain, while conscious perceptions of these actions of the will are only possible if they are slightly in the past. This means that awareness of a decision, that is the possibility of thinking about it, could only come after the act of willing, when the act was no longer slightly in the future. In this way, a human could bridge a "gap" between the near future and the near past. This gap is in fact a form of the gap between causes and effects possessed by chaotic systems. A more detailed investigation is clearly required to see whether the explanation given here works. Let it be noted finally in this connection that Roger Penrose mentions in "Shadows of the Mind" as a possibility a similar explanation for the psychological experiments discussed here, involving a spread in time of consciousness.

To what extent the same situation could exist for other living organisms, is a more open question. It is not so clear whether corresponding beings extended in time acting on various chaotic processes, would need to have the same structure in time as that which is proposed here for humans. If the time structure is similar, the nature of what is associated with life, the "etheric" in certain occult traditions, might be closely associated with the nature of time.

Another aspect may also be mentioned before the end of this section. Human society is very complex and certainly chaotic and unpredictable, with a strong dependence on interactions between individuals, though this is difficult to prove mathematically. This is clear if we examine history and wonder how a very small event might have completely changed it. For instance, if an accident had badly injured the nineteenth century German statesman Bismarck in early childhood, Germany might not have needed a war with France in order to attain unification (the Franco-Prussian war of 1870), which led to the loss of Alsace-Lorraine by France, French desires for revenge, the First and Second World Wars and the tragic events which followed them. The idea of a very small event changing the course of history has been used in fiction. Smaller groups of people similarly may also be influenced by almost imperceptible events. Even the stock exchange appears to be sensitive to very small perturbations, which unfortunately can have an extremely large effect on present-day economies. Benoit Mandelbrot, who invented the expression "fractal" for the geometry associated with chaos, has emphasized the role of chaos in the mathematical description of what happens at the stock exchange. It is for such reasons that we may suppose that various sorts of other beings, who are not directly perceived by the senses, may be able to act within groups of people. In fact, Rudolf Steiner, like other spiritual teachers, speaks of both beneficial and harmful invisible beings which act in the world, including some which act in human society. Rudolf Steiner uses the names of angels and archangels, which are also known to Judeo-Christian tradition. The presence of such invisible beings is evident in rather subtle ways.

3. *Is there a soul content in the domain of pure ideas?*

Reasons have just been given suggesting the possible existence of beings which have a component extended in time. We may wonder whether beings can also exist which are completely outside time. The existence of such beings might be expected to be related to a behavior which other beings could neither completely predict nor completely control. In fact as we saw at the end of the last chapter, a sort of "indeterminacy" exists even in mathematics, in the realm of eternal ideas outside time, thus at least suggesting the existence there of separate beings! The great monotheistic religions state that God is outside time; the question is whether we can see real indications for the existence of other beings outside time. In addition would such eternal beings possess something corresponding to the three soul abilities?

This question can be stated in another way. Indications have been found for the presence of soul in physics, that is of Popper's and Penrose's second world of mental experience within their first world of physical states. Do there then also exist indications for the presence of soul in their third world, that is in that of Penrose's third world of pure ideas, which includes the ideas of mathematics?

Let us start our present considerations with the relationship between the mathematician and the mathematical concepts which he or she studies. What is involved is the relationship of a human being with a part of the world of eternal ideas. The mathematician can discover something about certain aspects of mathematics, but if he or she tries to apply the discovery to prove something else, the mathematical contents or "objects" studied will "resist" this being done. In fact, certain mathematical statements such as Fermat's last theorem mentioned in chapter 3, are very simple. If such a theorem is stated as a hypothesis, one does not have to be a great mathematician to understand it; however the proof can be extremely difficult. There is, as we saw, no standard mechanical procedure to prove all theorems, or even to know whether certain theorems are provable. In order to prove a theorem a mathematician must know about other relevant theorems as well as having the ability to act in the world of mathematical ideas, that is to use will power. Once the theorem is proved, its truth is known in addition to previously proved theorems, and the mathematician is satisfied. Again contentment or happiness equals knowledge times an ability to act. Therefore the mathematician needs soul abilities in order to overcome the resistance of the objects of mathematics. This implies that mathematical ideas, including numbers, are real and not just a human invention. The very great, if not the greatest of Greek philosophers, Plato, considered pure ideas to be real and what is perceived by the senses to be imperfect reflections of these ideas. Many if not most mathematicians (but very few contemporary philosophers of mathematics) are "Platonists"; they believe in the reality of what they study.

The reasoning used up to now does not however directly throw light on the question as to whether the pure ideas of mathematics themselves possess something corresponding to the soul abilities, nor even whether they actually resist each other. This question is less easy to investigate than that of the relation of a mathematician to mathematics. What can be stated, however, is that the fact that it can be impossible to prove a theorem from any other known theorem, without one even being able to prove this characteristic, as we saw in section 7 of chapter 3, appears to indicate a great "distance" between different aspects of mathematics which contain the proofs of different theorems. A mathematical statement is not "dominated" by other "far away" statements. As noted in that section, at one time people wondered whether Fermat's last theorem was provable. One might think of such a "distance" and "lack of domination" as revealing a kind of resistance between these different aspects of mathematics. It is then this resistance which must be overcome by the will of the mathematician, who wishes to use what he or she already knows about certain aspects of mathematics to prove something which can only be done by understanding other aspects of mathematics. What could be the soul content of what then appears to be a resistance between different aspects of mathematics and of associated beings, needs however to be investigated further. Let it be noted in this connection that various soul qualities have been associated with numbers in spiritual traditions at least since the time of the ancient Greek teacher Pythagoras. The book written by the author's father mentioned in the preface is a more modern approach to this concept.

The ideas of mathematics are of course not the only eternal ideas which exist. Abstract words such as, good, beauty, love, wisdom, freedom, happiness, desire, selflessness and selfishness, represent ideas which are very hard to define. In fact a dictionary will define a word in terms of other words and it is possible by looking at the definition of the words used in the definition of the original word to find again the original word! In this way the attempt to understand the meaning of a word can lead one around in circles. In any case understanding what an abstract word may mean, what the ideas behind it are, is a task for the philosopher; such ideas will resist being understood as the ideas and theorems of mathematics resist a mathematician. This again suggests the presence of conscious beings who are connected with ideas of this kind.

The question of whether universal ideas or "universals" exist independently of individual objects (or the bodies of living things when living organisms are considered) whose nature and properties are describable using universals, was the subject of a great philosophical debate in Europe during the middle ages. An example of this type of question is: does happiness exist by itself or only because individual human beings and animals are happy? The "realists" believed in the real existence of universals independently of individual objects, while for the nominalists following Roscelin, they were only words.

The contents of the third world of Penrose can be seen to correspond to the spirit and to one or more eternal beings of the spirit. This world would be a world where the basic constituents of the other two worlds or their archetypes exist. We may expect that at least many of these constituents or rather separate beings are far beyond what can be grasped by ordinary present-day human thinking. Such archetypes were perceived by the spiritual experiences of, for example, Rudolf Steiner. In this framework the ideas of mathematics may be thought to contain the least "indeterminacy". Mathematical proof works in very many situations; it is only in the twentieth century that limits have been found to the power of mathematics. This is similar to what happens in physics; many properties and events are predictable in what on the human scale appears to be the almost "dead" processes of physics. It is therefore not surprising that almost "dead" mathematics with relatively little indeterminacy, was very successfully applied to almost "dead" physics, which has also relatively little indeterminacy. The relation of mathematics—and the discovery of new concepts in general—to the spirit and to spiritual teachings can be seen in another way. Great discoveries tend to be made following a certain sequence of events (described in the already mentioned book "The Quark and the Jaguar" by Murray Gell-mann). The nineteenth century German physicist Herman von Helmholtz stated that the birth of a new idea needed three stages—saturation, incubation and enlightenment. Murray Gell-mann describes the first stage as one of filling oneself with the difficulty of a problem and making attempts to overcome it. During the second stage conscious thinking becomes useless, even though the problem is still within oneself all the time. One might think that in the second stage the problem is still tackled unconsciously, or at least according to the interpretation of the unconscious given in the previous section of this chapter, by other beings with whom the human being is in contact. In the third stage a solution suddenly appears, while one can be doing something which is completely different. This is a sudden enlightenment or illumination. The French mathematician Henri Poincaré added a fourth stage, that of verifying that the solution really was correct. Another French mathematician, J. Hadamard, studied how mathematical discoveries were made.

This process in the discovery of new concepts, including those of mathematics, can be compared with the process of spiritual development of a human being who wishes to experience spiritual worlds, which cannot be perceived through the senses, as described by Rudolf Steiner in the book "An Outline of Esoteric Science" (Anthroposophic Press 1997). He describes several stages of perception of "invisible worlds", leading to what is really spiritual perception. These stages are "imagination", "inspiration" and "intuition". The stage of "imagination" (not to be confused with "imagining" unreal things) is reached through various exercises involving the formation of mental images of things not perceived by the senses, followed by concentration on these images. In this stage spiritual realities appear through living images, but are not directly perceived. The stage of inspiration where spiritual realities are perceived is attained by making the images disappear, while retaining only the effort made to produce these images. In that stage truths appear immediately without it being necessary to think about what is observed. The following stage of intuition, that is of direct contact with spiritual beings, is also reached through the effort in making the images of imagination disappear.

It might appear that the process of performing exercises to reach the stage of imagination corresponds to the filling of oneself with a scientific or mathematical problem, or what Helmholtz called saturation. The "inspiration", or perhaps even the "intuition" of the solution, follows a stage when awareness of the problem disappears from consciousness. Poincaré's fourth stage might then be related to living in a world of spiritual realities. However, we must remember that the spiritual experience described by spiritual teachers like Rudolf Steiner goes far beyond the processes of scientific and mathematical discovery. As we have mentioned, spiritual archetypes may be far beyond what can be grasped by ordinary present-day human thinking.

4. Comments on the concept of "energy"

At this point it may be useful to consider in more detail the concept of energy. Its role in pre-twentieth century physics was discussed in section 2 of chapter 2. The word "energy" is often used by certain spiritual movements and is frequently believed to be directly connected to spiritual realities. Some speak of different forms of energy including "spiritual" energy. Unorthodox types of medicine speak in addition about "energies of the body". We shall look at the concept of energy in physics whereby we can see how much of what is said on this subject is the result of unclear thinking and is in some ways materialistic.

Let us start by looking at the origin of the word. According to the French "Robert" etymological dictionary ("Dictionaire de l'etymologie du francais" by Jacqueline Picoche, 1973), the word energy (energie in French) is derived from an indo-european root which means to act. This root "werg" or "worg" is at the origin of the English word "work" and the German word "werk". In the development which led to the word "energy", we find the Greek word "ergein", which means an acting force. For the great Greek philosopher Aristotle "ergein" was contrasted with "dynamis", which is a possibility of action. We can therefore see that the word energy is derived from the idea of action which, we should remember from our previous discussions, is not the only defining aspect of the world.

In elementary physics a fixed amount of energy can, in principle, be converted to mechanical "work", that is, enable an object to move against a force which is acting on it. In fact there exist many forms of energy in physics such as light, heat and electrical energy. The principle of the "conservation of energy", mentioned in chapter 2, states that one form of energy can in principle (not always in practice) be converted to another form; a fixed quantity of one form is convertible to a fixed quantity of the other. Einstein's famous equation $E=mc^2$, mentioned in chapter 3, states that even the mass of a body can be converted to other forms of energy. The energy of physics is the ability to act in the realm of physics or, in the words of the "Encyclopedic Dictionary of Physics" (Pergamon Press 1961), "energy may be defined as the capacity for producing effects".

However, the perception of different forms of energy, that is the inner experience of them by human beings, is not at all the same. Though the conservation of energy is accurate when numerical measurements of energy are made, it does not describe the human experiences associated with different forms of energy.

The total quantity of energy available in a certain form may often be the appropriate ability to act in the type of physics relevant to the human scale, as mentioned in section 2 of this chapter, where a possible action of beings extended in time on chaotic systems was considered. However we saw that this is not the case in situations where the Heisenberg indeterminacy principle needs to be taken into consideration; there in the sphere of quantum physics it is the accuracy with which a fixed quantity of energy can be applied, which appears to be equal to the ability to act. It is in such ways that the ability to act need not be the same in different situations. The ability to act is almost certainly not the available quantity of physical energy, nor the accuracy with which this energy is applied when a human body is cured of an illness. It is therefore better not to call all forms of ability to act "energy".

The role of the total quantity of energy for conscious beings acting on physical chaotic systems can be compared with that of something which intervenes when human beings have relations with each other. Money plays a similar role to that of energy in relations between people and groups of people!. Its possession gives human beings the ability to act. Money can also be expressed in the forms of different currencies used in different countries, each of which can be converted to another through an exchange rate. There is then something corresponding to the conservation of energy when one currency is converted to another one, though the rate of conversion, that is the exchange rate, is not constant. In any case different currencies are not equivalent; the relative prices of different products are not the same in different countries, these prices depending on the economy and culture of a particular country. This reminds us of the fact that human perceptions of different forms of energy are not the same. This suggests that energy is not necessarily spiritual; in certain physical situations it plays the role played by money in human society. I must admit that I am tempted to connect the present excessive concern with energy in certain circles with the excessive preocupation with money at the present time in most of the world. Indeed certain individuals who pretend to be "gurus" are especially interested in the money they earn through their activities.

5. Is there a kind of trinity behind the various conscious beings?

There is a further step which may be taken if many different sorts of conscious beings exist. The idea of the existence of a divine trinity behind the phenomena of the world is present in more than one culture. Hinduism speaks of three supreme gods: Brahma, who creates, Vishnu who maintains and Shiva who destroys. We should note that such a concept is closely related to the time trinity of past, present and future, although it is not quite the same. The creation of an object which exists in the present occured in the past. The object was maintained in the past after being created, is being maintained in the present and will be maintained in the future, until it is destroyed. However, three gods of this sort cannot be thought of as existing separately in the world of eternal ideas, which is neither created nor destroyed.

The Greek philosopher Heraclitus, mentioned in the last section of chapter 2, asserted that there were three fundamental principles in the world. They were "Theon", a kind of divinity, the "Logos", which can be thought of as the words or language of the Universe and of philosophers, while the last principle was perpetual change. It may be noted that the English word "logic" is based on the idea of the Logos. A few centuries after Heraclitus, the Logos was a basic concept in stoic philosophy. The author of St. John's gospel quoted the same principle of Heraclitus in the prologue at the start of the gospel, saying that the Logos or word, which was in the beginning, was made flesh and lived among human beings in the form of Jesus Christ. Thus one of the basic doctrines of Christianity is related to what had been previously taught by Heraclitus. It might be possible to extrapolate such an interpretation; in that case, Theon would represent the Father and perpetual change the Holy Spirit. It seems however that Heraclitus was influenced, at least to a certain extent, by teachings of the Greek mysteries. His three principles might indeed also be related to Rudolf Steiner's stages of spiritual perception, briefly described in section 3 of this chapter. Perpetual change is then the basic property of the living images of the stage of imagination, while the direct perception of truth is attained in the stage of inspiration. Direct contact with spiritual beings in the stage of intuition can then be related to perception of the divine. It may therefore even be possible to connect Rudolf Steiner's stages of perception with the Trinity.

It is perhaps dangerous to try to relate a divine trinity to what may be discovered through scientific research. There is a story about the great teacher Alain de Lille who was called the "Doctor Universalis" and who lived in the twelfth century. He was walking along the banks of the Seine in Paris in 1168, the day before he was due to talk about the Trinity, when he saw a small boy trying to make all the water of the Seine enter a small hole. Alain de Lille told the boy that this was impossible, whereupon the boy replied that it was also impossible to speak about the Trinity. According to the story, Alain de Lille was so shocked that he abandoned teaching the following day.

In order to proceed further, we shall again look at quantum physics. We saw in chapter 3 that quantum physics can only describe the different possibilities of what may happen; one of the possibilities is realized by "wave function collapse". It is in this way that the occurrence of real events in time on the human scale is related to the quantum world. We saw that this phenomenon is now understood by many physicists through the interaction of small systems obeying the laws of quantum physics with a very large "environment", that is, with the outside world, which is indeed virtually infinite when compared with the quantum system. This process is usually extremely rapid. We can think of each interaction or meeting of all the small quantum systems which exist with particles of the environment as being a kind of "death" process of a host of possibilities possessed by the beings associated with quantum physics, whose existence was proposed in the first section of this chapter. This is very similar to the way the human experience of the present was described as a sort of death of the possibilities of acting on the future in the last section of chapter 2. In interactions the particle-like properties of the components of a quantum system should be taken into account, the particle needs to be in a certain region of space near another particle if it is to interact. After an interaction with the environment many different possibilities again appear, almost all of which will again die in the next interaction. It is in such a way that interaction with the environment makes us think of a continual process of birth, followed by death associated with confinement to a small region of space, which is followed in its turn by re-birth. The world on the human scale then appears to make the quantum world of "sub-nature" die when it interacts with the human world.

Interaction or meetings occur when the path of a particle with quantum properties crosses the paths of other particles belonging to the environment. It is possible to picture this through Christian symbolism. We may view the death process of the quantum world as a process of continual "crucifixion", as a quantum system is crossed by something belonging to the environment. Beings of sub-nature are crossed or met by the large scale world. In this way real events can be produced in time and the rhythm of the succession of these events is a language or "logos". The original world of possibilities is that of the Father, death is connected with the Son and the re-birth of possibilities with the Holy Spirit. What is associated with the Holy Spirit then plays the role of Father in later interactions. This can be directly connected with the spiritual stream of the Rosecrucians and the way they related to the Trinity, involving birth in the Father, death in Christ and re-birth in the Holy Spirit:

Ex Deo nascimur
In Christo Morimur
Per Spiritum Sanctum reviscimus

In section 2 of this chapter it was proposed that beings other than those connected with the Heisenberg uncertainty principle act in the large scale world. Such beings are also associated with what is unpredictable. In this case when two such beings meet the result of the meeting must be unpredictable. Only one of the many possibilities of interaction between them will become real; the other possibilities die. Death is then a necessary part of the realization of possibilities in the partly unpredictable universe in which we live. Without death there could be no re-birth. The Son or cosmic Christ may, following what has been said, be seen to be present in meetings. This way of understanding what happens in the universe need not and indeed does not, at first sight, suggest the existence of an all-powerful God.

It may be possible to conceive of the Trinity as consisting of three (or one) independent consciousness(es), which manifest themselves (itself) through what is accomplished by all the different conscious beings of the universe, without controlling all of what is performed by those other conscious beings.

What has been said here can be stated in another way through Rudolf Steiner's "Foundation Stone" meditation. This is the basic meditation for members of the Anthroposophical Society, founded by him. The idea of the Trinity of Father, Son and Holy Spirit is fundamental in this meditation. The second part is directy connected with the idea of the environment, though the meditation was given to members of the Anthroposophical Society in 1923, many decades before physicists thought of linking the environment to quantum physics:

For the will of Christ rules in the periphery,
Blessing the soul in cosmic rhythms.
Denn es waltet der Christus-Wille im Umkreis
In den Weltenrhythmen Seelen-begnadend.

The human being, unlike lower beings, is self-conscious; he or she can say "I am". In this way it is possible for a human to meet his or her inner self. It is in such a manner that the meeting with the inner self may also be connected with the Son or Christ. Further consideration of this question is outside the scope of this book.

Let it be emphasized that the role of the Trinity proposed here does not "prove" the doctrines and dogmas of the different churches concerning the historical incarnation of Christ. In any case the aim here is not to make the reader believe any particular doctrine or dogma and even less to make him or her submit to the hierarchy of a church.

5

Towards a New Science

1. *How a threshold appears to have been crossed*

The striking results of twentieth century science, described in chapter 3 of this book, radically changed the way scientists think about the world. We have seen in the last chapter how such results can be understood by the presence of conscious beings possessing the soul abilities of knowledge, happiness and the ability to act. Therefore, in spite of the basic assumptions made at the birth of modern science, it would appear that soul cannot be completely eliminated from science. This means that the world of physics can be thought of as being a kind of illusion, or to use an eastern expression, "maya", hiding the existence of soul.

In particular the path of physics whose development was determined by its basic assumptions led to certain logical conclusions in quantum physics and the discovery of the strange phenomena associated with it. As we saw in Chapter 4, these phenomena can be at least partially explained by the presence of conscious beings who resist each other and for whom happiness cannot be greater than a constant of physics. Such beings appear to belong to an amoral world of "sub-nature". Other scientific results appear to be understandable by the simultaneous presence of quite different sorts of conscious beings in chaotic systems and even in the world of pure ideas. Reasons for also believing in the presence of a divine Trinity behind beings of different kinds were also given at the end of chapter 4. However, the author of this book does not claim to have rigorously "proven" these statements; what can be said at least is that they help us to understand many aspects of the world, including the inner experiences of human beings, as well as having a certain intrinsic logic.

In our approach we use to a large extent human inner experiences as a guide to understand nature. A human being has consciousness, soul abilities and many sorts of experience, which cannot be reduced to the concepts of physics based on space and the space-like aspects of time. It is not supposed here that other beings are like humans, which would be an extremely dubious assumption to make, but that human inner experience reflects fundamental properties of the world.

We may at this point recall the debates described in chapter 1, about whether science is only social relations. It is now possible for us to give a more precise reply than given in that chapter. According to what has been said in this book, the social relations are with other beings, that is with beings who may be other humans, as well as with those of nature, of sub-nature or in the world of pure ideas. Our relations with other humans and the type of society we live in will clearly influence our relations with other kinds of beings and so influence our science and technology. Therefore we cannot dissociate science and its history from other aspects of human society and history. Some of those other aspects will also be examined in this chapter.

If we look at the discoveries described in chapter 3, it appears that around the beginning of the twentieth century, science crossed a kind of "threshold" into realms of experience which are very different from those of the normal physical world of everyday life. The descent into worlds of sub-nature was especially noteworthy; if such worlds are understood properly, it appears that they can teach us many basic lessons. It may be possible to view this realm of sub-nature, or Ahriman, as possessing certain special, even spiritual, characteristics. This means that it would have been difficult to learn such lessons if science had not been concerned with sub-nature. However, what has been said in this book indicates that to learn them we must not mistake sub-nature for a kind of divine world.

The nature of the threshold becomes clearer if we look at certain dates:

Experiment of Michelson and Morley: 1887
Discovery of radioactivity by Bequerel: 1896
Properties of electron determined by Thomson: 1897
Planck's quantum theory: 1900
Einstein's theory of the photo-electric effect: 1905
Einstein's theory of special relativity: 1905
Rutherford's model of the atom: 1911
Bohr's model of the atom: 1913
Einstein's General theory of relativity: 1915
De Broglie proposes matter waves: 1923
Schrödinger's wave mechanics: 1926
Heisenberg's indeterminacy principle: 1927
Gödel's theorem: 1931
Realization of importance of chaos: the 60's

It may be relevant to compare these dates with certain traditions concerning the spiritual evolution of mankind, according to which great changes had to occur around the beginning of the twentieth century. For instance, according to a statement made in the late nineteenth century by Blavatsky, the principal founder of the Theosophical Society, in her book "The Secret Doctrine", mankind was about to enter a new cycle in its development. In this connection there is an Indian tradition about different sorts of ages or "Yugas". The four yugas are Krita Yuga, Treta Yuga, Dvapara Yuga and Kali Yuga and may correspond at least to some extent to the four ages (golden, silver, bronze and iron) of the ancient Greek poet Hesiod. According to such Indian traditions, mankind entered "Kali Yuga", an age of conflicts and darkness, in 3002 BC; certain teachers state that it would last about 5000 years. Rudolf Steiner said that it ended in 1899 and that following its end, mankind would progressively acquire a higher form of the spiritual abilities which had been previously lost. The French student of traditional Indian teachings Alain Daniélou in "Le Destin du Monde d'après la Tradition Shivaite" (Espaces Libres, Albin Michel 1992) states that Kali Yuga ended in 1939, but that it is being followed by a "dusk" of Kali Yuga", which would end in 2442 with the almost total destruction of mankind as it exists at present. However, to be fair, there is some disagreement about the length of Kali Yuga; in his article "Quelques Réflexions sur les Cycles de l'Histoire Humaine" in the French journal "Troisième Millénaire" (page 52 of Nr. 41) Jean—Louis Siemens states that according to the traditional "Mânavadharma Shâstra" (the laws of Manu), Kali Yuga should last 432,000 years, which means that 427,000 years of Kali Yuga still remain to be experienced!

In any case, we do not need to rely on ancient teachings to see that mankind has crossed a threshold. As we saw in chapter 1, science plays an essential role in the modern world. In addition, the scientific discoveries of the twentieth century are not the only signs of such an event. Modern technology, which is often based on present-day physics, enables people to have "sham" spiritual experience. It is possible to produce artificial images on screens such as those of the cinema, television and computers, using (except for the cinema) electronics which require the use of twentieth century physics of the very small. Systems of virtual reality are much more powerful because they enable people to have different kinds of artificially produced sense impressions simultaneously, produced so as to correspond to the results of computer calculations. There are many practical applications of virtual reality. It is possible, for instance, taking only one of a multitude of examples, to simulate perceptions of a house which is not yet built, so as to see what living in it would really be like. Such experiences of disconnection from the nor-

mal perceptions of the body are parodies of spiritual experiences following meditation, which are also associated with disconnection from body perceptions.
Indeed the dangers resulting from such disconnection are not completely dissim-
ilar. In particular care should be taken that mental health is not harmed. Present
methods of producing virtual reality are still rather primitive; far more refined
approaches, perhaps through the production of perceptions by direct action on
the nervous system, can be envisaged. One way of understanding such developments would be to think of them as an attempt to render true spiritual experiences impossible, bearing in mind the birth of new spiritual abilities at the
present time announced by various teachings, such as those of Rudolf Steiner.

Though rather beyond the scope of this book, a brief look at twentieth century history seems also to suggest the crossing of a threshold. This history has
involved many tragedies, though that is not the most original feature of the twentieth century. It is true that great massacres happened in the past, during the
Mongol invasions of the thirteenth century, for example. The arrival of Europe-
ans in America was in many ways a disaster for the people already living there.
However, in addition to its tragedies, the twentieth century has seen attempts to
create large scale organized societies of a kind which never existed before in the
physical world. Communism, in spite of its claims to be scientific, was to a certain extent based on utopian ideas. The book "Utopia" (from the Greek meaning
nowhere) about an imaginary perfect Communist society, was written by Thomas More in the early sixteenth century. The society described in the book has
certain totalitarian characteristics, such as the impossibility for the inhabitants of
Utopia to have privacy. Thomas More was a very religious and moral man.
According to Rudolf Steiner in a lecture given on May 2, 1916, More was,
through his meditations, able to have spiritual experiences in his sleep which people normally do not have, but which he was unable to communicate consciously.
A description of these experiences was related in "Utopia"; Therefore, we can
think of the utopian ideas of Communism as being an attempt to transplant spiritual experiences into the physical world of everyday life, in spite of the extreme
materialism consciously believed in by Communists. Following the industrial
revolution and the sufferings of working people due to it, a form of utopia
seemed to be very attractive to large numbers of people. Utopias are, however,
out of place in the physical world and it was only possible to attempt forms of
Utopia in it using extreme violence. Spiritual experiences obtained in such a way
are illegitimate. The violence and repression which reached a peak under Stalin
replaced the ideals; rotten societies were produced, most of which had collapsed
before the time of writing of this book.

The dangers of wrong sorts of spiritual development were mentioned at the end of chapter 2 in connection with the discussion about time. According to Rudolf Steiner, someone who wishes to have spiritual experiences must cross a kind of "abyss". Some of the tragic events of the twentieth century can therefore be viewed as connected with the dangers associated with wrong approaches to the spiritual and the presence of such an abyss.

2. How might it be possible to study the phenomena of nature from physics to biology in a new way?

In this book soul content has been looked for in partly unpredictable phenomena and in finding certain features belonging to them, corresponding to the soul aspects of knowledge, happiness and the ability to act of various conscious beings. What should be emphasized at this point is that even if a phenomenon has three aspects, this does not by itself prove the presence of consciousness with a soul content. Soul content may, however, at least be indicated if something partly unpredictable is present and if a connection can be found between each of the three aspects of the phenomenon and one of the soul aspects of knowledge, happiness and the ability to act. These conditions are not always satisfied, so claims sometimes made in anthroposophical circles for a relation between all sorts of phenomena and the human abilities of thinking, feeling and willing, are often not easy to justify. However, we can expect that real indications of soul aspects belonging to conscious beings may be found in many situations and phenomena, not studied in this book. Such indications need to be looked for and investigated in detail, using the inner experiences of human beings as a guide. Indeed the direct study of inner experience and perceptions (such as our study of time) can teach us a lot. Let us in this connection recall Goethe's study of colours, described in chapter 2. The physics of the very small needs to be investigated in much more detail than done here.

In addition, it should be emphasized that this book has examined time rather than space. This is because of the fact that time can be more directly related to soul qualities. It should however be possible to see soul aspects in space also, at least because various forms of spatial separation can be thought of as isolating the fields of activity of different beings from each other. There are also ways of looking at space that are different than those generally practiced in physics and which have particularly interested people seeking another type of science based on the teachings of Rudolf Steiner, that is, in the framework of Anthroposophy. To briefly summarize this type of approach, it must be pointed out that certain sorts of geometry exist in which distances are not directly considered, but rather special properties derived from distance. Properties called "cross ratios" are unchanged if the distances and angles are transformed in various precise ways. This is the case with Projective Geometry, which has a very important feature—that points and planes (flat surfaces) are equivalent; each property of a plane is accompanied by a corresponding property of a point. Thinking about projective geometry can for this reason help to free us from the idea that space must be always thought of as made up of points; it can equally be thought of as being made up of planes. Projective geometry is described in "Projective geometry" by Lawrence Edwards (Rudolf Steiner Institute 1985). Geometries exist which are "between" the geometry of distance as experienced normally by people in everyday life, that is, "Euclidean" geometry and projective geometry. There also exists a geometry of what is called "counter space" studied by George Adams and Louis Locher-Ernst, which is, in a way, "opposite" to Euclidean geometry and its use of distance, the role of points in normally perceived geometry being replaced by that of planes. Finally there also exist geometries with properties between those of the last mentioned type of geometry and those of projective geometry.

If we try to relate these geometrical considerations to physics and to other sciences, we can firstly note that the relation between points and planes reminds us of the roles of particles and waves in quantum physics. A particle is a kind of "expanded point", while the surfaces of waves in three dimensional space can become almost plane. Georg Unger in "Forming concepts in physics" (Parker Courtney Press, Chestnut Ridge, New York, 1995), hints of a possible application of the simultaneous use of point-like and plane-like models to quantum physics. Using such means, the soul characteristics of this world may be describable geometrically. Nick Thomas, in an article "Rethinking physics" ("Newsletter Articles Supplement" Nr. 2 of the science group of the Anthroposophical Society in Great Britain, p. 1, 1996) and more recently in the book "Science between space and counterspace. Exploring the significance of negative space" (Temple Lodge Publishing, London, 1999) goes further, in describing a detailed attempt to refound physics based on these geometrical concepts. He mentions a relation between counter-space (as well as of the geometries with properties between those of the geometry of counter-space and those of projective geometry) and what is called the "etheric" in Anthroposophy. The etheric is especially connected with life. Rudolf Steiner described four "ethers", 3 of which Nick Thomas considers as being described by one of those types of geometry. There is, according to Nick Thomas, a similar relation between Euclidean geometry and the geometries between it and projective geometry and the solid, liquid and gaseous states of matter. In addition, objects in physics simultaneously fill both the ordinary space of Euclid and counter-space, which have contradictory properties, leading to strain in one space followed by stress and finally to a force. He is able to derive certain laws of physics using this approach. It remains to be seen how successful Nick Thomas will be in the future in rethinking the whole of physics in this way.

As we saw in chapter 3, chaotic systems are described by the geometry of fractals. It may, in fact, be possible to relate the sorts of chaos in living organisms to the types of geometries just mentioned. Counter space can indeed be defined in an infinite number of ways with respect to the space of Euclid, because each point of Euclidean space can be defined as being infinitely far away in a particular geometry of counter space. It is for this reason that the different possible geometries of counter space, when considered together, may show fractal properties, with infinite structure at each point of Euclidean space. We can now mention the work of George Adams and Olive Wicher on plants, described in "The living plant and the science of physical and ethereal spaces" (Goethean Science Foundation, Clent, Stourbridge, Worcestershire, England 1949). They found that many aspects of the forms of plants could be understood by the geometry of counter space with its infinitely distant point. As they indicate, many such points may be present for the same plant. We might perhaps think of the counter space description of a plant as corresponding to the wavelike aspect of phenomena in quantum physics, while what in a plant is best described by Euclidean geometry would then correspond to the particle-like aspects of quantum physics.

Geometrical studies have also been performed by Lawrence Edwards as described in "The field of form" (Floris books 1982). He studies what are called "path curves", which are defined by certain transformations of structures permitted by projective geometry. He finds that various shapes of living organisms, including plants and the heart, can be well described using path curves. It may be possible to relate this kind of work to our preceding considerations about the geometry of chaos present in living organisms, but it remains to be seen whether such connections can be made.

Other sorts of future of research can be indicated if we bear in mind that chaotic systems are sensitive to minute effects, some of which might appear at first sight to lead to completely unexpected and even "crazy" phenomena from the point of view of present-day physics. For instance, Lawrence Edwards finds that plant shapes are influenced by the positions of the planets; if confirmed, such a result would appear to be quite difficult to explain by normal physical forces. Researches have also been undertaken in anthroposophical circles to detect effects of the etheric in certain experimental situations, such as that of substances coming from living organisms by the crystallization of copper chloride (see "Sensitive Crystallization Processes. A Demonstration of Formative Forces in the Blood" by Ehrenfried Pfeiffer, Anthroposophic Press, Spring Valley, New York, 1975). A solution of copper chloride to which a small quantity of that substance has been added, evaporates, leading to the formation of crystals, having in each particular situation a characteristic pattern. Chaos appears in fact to be present in crystallization; its possible role in this kind of experiment needs elucidation.

Some of the phenomena studied in psychic research might be due to similar effects, when they are real and not due to cheating by those involved in trying to convince others of the existence of such phenomena, such as in the phenomenon called "psychokinesis". In psychokinesis the mind is supposed to act directly on matter outside the body in a "non-physical" way. The possibility of such action appears in fact to be not completely unreasonable, in view of our discussion in section 2 of the last chapter on how a human being might be able to control his or her own body when chaotic effects occur.

The author of this book has in a somewhat similar approach looked for "crazy" astronomical facts, which should not occur according to present ideas about the cosmos. Such facts, if significant, would be dependent on observations being made from the earth, which is the home of Man and would therefore suggest that Man has a certain significance in the cosmos. The existence of certain facts of this sort is actually indicated, including, for instance, relations between the positions of certain bright stars in the sky as seen from the earth. The position of Sirius, the brightest star for the human eye, is near 90° (88.73°) from the third brightest, αCentauri, which, according to astronomical methods of distance determination, appears to have about half the distance of Sirius. When studied with telescopes αCentauri is seen to be actually a system of three very close stars, which are the nearest stars to the solar system. Similar relationships exist between the positions in the sky of observed novae which are bright as seen from the earth, these being among the objects which I study in my normal astrophysical work. Novae typically brighten rapidly by a factor of more than 10 0000, before fading usually much more slowly to a brightness close to the brightness before the outburst. An attempt was even made by me to predict the sky position of future novae, which has not been successful till now.

A different example of an astronomical fact related to the earth is the weak electromagnetic radiation coming from all directions of the sky, whose existence was mentioned towards the end of section 2 of chapter 3 (devoted to relativity). It was stated there that a first deviation from "isotropy" (equal amounts of radiation coming from all directions), is usually interpreted as produced by an absolute motion of the solar system with respect to this radiation. The velocity of this motion, which has been measured as being 369.5 km/s from observations with the COBE satellite, can be simply related to two other basic velocities. The first is the speed of light of 299 790 km/s, which plays a fundamental role in relativity. The other velocity is what might be considered as being the speed of the "most basic" motion of the earth, which unlike other motions is not defined with respect to any other single astronomical body, that is, the speed of its rotation at the equator, equal to 0.46510 km/s. Now the ratio of the speed of light to the absolute speed of the solar system with respect to the radiation coming from all directions in the sky, is very close to the ratio of this absolute speed to the speed of the earth's rotation at the equator. In fact, the reader can verify that if one of these ratios is divided by the other, a value very close to 1 is obtained: 1.0213. Objections can be raised, however, against such a search for "crazy" astronomical facts. If one plays with a large enough quantity of different numbers, one will eventually find apparent relationships between some of the numbers, simply because of the laws of probability. There is always a certain probability of two numbers being almost the same, without any other special connection existing between them. It is difficult to eliminate such a possibility for the types of relationships mentioned. In addition, if one looks at the positions of those stars in the sky which appear to be the brightest for the human eye as seen from the earth, one must remember that these positions are not fixed, but slowly change with time, making the significance of anything based on the present positions doubtful, unless the present time is considered to be specially significant. The sky seen in a few thousand years will be different, perhaps with other relationships between the positions of the brightest stars. We can therefore say that the situation concerning the significance of this type of astronomical relationship is somewhat uncertain and it is not clear to me to what extent I can conclusively prove something in this way. However, it might not be useless to continue looking for similar simple striking types of relationships which, like some of those already found, are obtainable without too much playing with numbers. Many significant facts may exist which not been searched for till now, because they would have been considered to be completely impossible.

3. New possibilities in sciences concerned with human beings and finding a better structure for society

The generally accepted "scientific" methods of studying human behaviour can appear very doubtful to a physical scientist like myself. Much seems to be based on statistics, without it being very clear what fundamental phenomena are really responsible for the various statistical results obtained. In addition, I sometimes wonder whether people working in such fields do not often do much more than invent complicated words, so as to dazzle the non-specialist. The approach of this book, based on the presence of different sorts of beings possessing knowledge, happiness and ability to act, may help to overcome such problems. Human beings do not only, according to this point of view, possess these three abilities, but are in addition influenced both as individuals and in groups by other human beings and by various kinds of non-human beings. In any case, it is clear that much research needs to be done in these fields in order to make such ideas more precise and find out how they can be applied.

One aspect of the social sciences can be linked with suggestions made by Rudolf Steiner about how to produce a more healthy society. These suggestions are found in particular in "Towards Social Renewal" (Rudolf Steiner Press, London, 1975). He advocates a threefold structure for a society based on a separation of cultural life from that aspect of society connected with the rights of each human (including laws and how they are decided on and enforced), as well as a separation of these two aspects of society from a third aspect of society connected with the economy. Culture is concerned with mental and spiritual life or more generally with the natural endowments of each human being. The rights of each human being belong to the world of human relations, while economic life is concerned with what is produced from nature. Rudolf Steiner states that the motto coming from the French revolution—Liberty, Equality, Fraternity—should apply separately to each of these three aspects of society. Freedom is what should rule the cultural life, equality that of rights, while fraternity should rule in the economy. In this way democracy is important in the life of rights, while solidarity between all human beings is important in the economy.

If we examine the three aspects of society, we can see that cultural life is related to what is sought by each individual in his or her "researches", which can be for instance in art and religious experience and even in practicing a hobby or succeeding in sport as well as a form of research in a field of science. Culture is in this way connected with kinds of "knowledge" of each of the members of a society. The aspect connected with rights involves how to make a society happy in its human relations, while the economy is concerned with the way a society acts particularly in its relations with nature.

Explaining the necessity for this type of social structure in a different way than done by Rudolf Steiner, we may say that, at the present time, human beings have become more and more conscious of themselves as separate autonomous individuals with different desires and tend to resist each other to an increasing extent. In fact we may state that something like the Heisenberg indeterminacy principle is more and more true in human relationships, especially in western society. Each human and group of humans tends to resist and fight others, that is, to try to both limit the amount of knowledge and the ability to act of others, in order to not be overwhelmed by them and so to be able to satisfy his, her or its desires as much as possible. In this way, limits are placed on the total "knowledge" of a society and on its ability to act, so the society as a whole becomes "unhappy". A separation between what corresponds in a society to knowledge, what corresponds to the feeling of contentment in human relations and what corresponds to the ability to act, can help to remove reasons for conflicts and so help to at least partly overcome this "unhappiness". In particular it should be possible to overcome the present domination of most of the world by the economy, which among other things limits the rights of human beings and manipulates culture through the media. Let us in this connection think of the scandal of large numbers of extremely poor people living at the present time in very rich western countries.

We may note that it is not an accident that this tendency of individuals to fight others increased at a similar time to that when physics discovered the Heisenberg indeterminacy principle. Phenomena were studied which "resembled" present-day human behaviour.

The threefold social order is not a utopia, like Communism, as I personally believed when I was younger, though it can at least partly lead to the realization of many social ideals. It cannot eliminate all conflicts, but can, if brought into being, improve things. However, it must be admitted that it is extremely difficult in the present world with, among other things, a global economy, even to start to realize a form of a threefold society; in fact much more work needs to be done than has be realized till now, before it can be seriously applied.

4. Closing Comments

I have in this book tried to show how it may be possible at least to start to bring soul and the existence of conscious beings into science. The presence of such beings may be indicated where indeterminacy occurs as well as in the presence of various kinds of "resistance". The soul experiences of such beings would then appear to be fundamental in understanding the nature of the universe. If the approach expounded here is justified, much more work needs to be done to see how fruitful it is. Such fruitfulness rather than any sort of "proof", which would appear at first sight to be nearly impossible, will be able to indicate how good the approach is. In this way it may be considered as a guide to future scientific investigations and not just pure philosophical speculation.

In order to create a science that includes soul, it will be necessary to overcome what seems to me a kind of fear of abandoning basic assumptions. Furthermore, the world of the human soul and its inner experiences, as well as that of the possible perception by it of physically invisible beings, is also that of dreams and nightmares. Bringing this type of world into a science like physics can indeed be frightening for many people.

Another science should also lead to other technologies and to other sorts of human society, which are more "ecological". Such new technologies might for instance be based on the collaboration with certain beings acting in chaotic situations. In any case, realizing the importance of inner soul experiences should lead to people being less willing to treat each other as machines, as sometimes appears to be case in certain biological experiments connected with human reproduction. However, we should have no illusions: an awareness of being always surrounded by other conscious beings will not necessarily make human beings behave any better. History, including recent history, has shown how cruel people can be to other people, not to mention how cruel they can be to animals. In this situation people will not necessarily be less cruel to any other sort of visible and invisible being, whose existence they may become aware of, unless they become afraid of the effects of such cruelty.

Science as we now understand it is less than five centuries old; it is still in its infancy or at best in its adolescence. The future will be able to bring into being a much more mature science.

Michael Friedjung was born in 1940 in England of Austrian refugee parents who had escaped from the Nazis. He was already deeply interested in science at eleven years of age, and uniting science and spirituality eventually became his aim. He studied astronomy, obtaining a Bsc in 1961 and his Phd in 1965. After short stays in South Africa and Canada, he went to France in 1967 on a post-doctoral fellowship and later was appointed to a permanent position at the French National Center for Scientific Research (CNRS) in 1969, where he is now Research Director. After living with the contradictions between official science and spiritual teachings, he began to see solutions to at least some of the problems, which are described in this book.

PUBLICATIONS QUOTED IN THIS BOOK

Adams, George and Wicher, Olive; "The Living Plant and the Science of Physical and Ethereal Spaces", 1949, Goethean Science Foundation, Clent, Stourbridge, Worcestershire, England

Barrow, John, D. and Tipler, Frank, J.; "The Anthropic Cosmological Principle", 1986, Clarendon Press, Oxford, England

Bitbol, Michel and Ruhnau, Eva; "Now, Time and Quantum Mechanics", 1994, Frontieres, Gif sur Yvette, France

Capra, Fritjof.; "The Tao of Physics". Wildwood House 1975 and Fontana 1976

Churchland, Paul, M. and Smith-Churchland, Patricia, "Could a Machine Think?", 1990, "Scientific American", vol 262, nr.1, p 26

Danielou, Alain; "Le Destin du Monde d'après la Tradition Shivaite". 1992, Espaces Libres, Albin, Michel, Paris, France

Edwards, Lawrence; "Projective Geometry", 1985. Rudolf Steiner, Institute, Phoenixville, Pennsylvania, USA

Edwards, Lawrence; "The Field of Form", 1982, Floris Books, Edinburgh, Scotland

Feschotte, Pierre; Les Illusionistes. Essai sur le Mensonge Scientifique", 1985, Editions de l'Aire, Lausanne, Switzerland

Feyerabend, Paul; "Against Method", 1975, New Left Books, London, England

Freeman, W. J., "The Physiology of Perception", 1991, "Scientific American", vol 264, Nr. 2. p 34

Friedjung, Michael, 1991, "Modern Physics and the Nature of the World", Network Newsletter ("Scientific and Medical Network"), Nr. 45, p 4

Friedjung, Michael, 1997, "Time: a Challenge to Physics", "Network" (Scientific and Medical Network), Nr.63, pg. 16

Friedjung, Walter; "Vom Symbolgehalt der Zahl". 1968, Europa Verlag, Vienna, Austria

Gell-mann, Murray, "The Quark and the Jaguar. Adventures in the Simple and the Complex". 1994, Little, Brown and Co.. London,

Gleick, James; "Chaos: Making a New Science", 1988, Sphere Books, London

Goldberger. A. L.; Rigney, D. R.; West, B. J.; "Chaos and Fractals in Human Physiology", 1990, "Scientific American", vol 262, no 2, p 34

Haroche, Serge,; "Entanglement, Decoherence and the Quantum/Classical Boundary", 1998, "Physics Today", vol 51, no 7, p 36

Holdrege, Craig; "A Question of Genes. Understanding Life in Context", 1996, "Floris Books, Edinburgh

Jeanière, Abel; "Les Presocratiques", 1996, Seuil, Paris, France

Kaufman, S.A.;"Antichaos and Adaption", 1991, "Scientific American", vol 265, no 2, p 64

Klein, Etienne; "Le Temps". 1995, Dominos, Flammarion, Paris

Kline, Morris; "Mathematics the Loss of Certainty", 1980, Oxford University Press

Masani, Sir Rustom, "Zoroastrianism: the Religion of the Good Life" 1962, Collier Books, New York, USA

Nottale, Laurent; "L'Espace-Temps Fractal", 1995. "Pour la Science", no 215, p 34

Penrose, Roger; "The Emperor's New Mind", 1989, Oxford University Press

Penrose, Roger; "Shadows of the Mind", 1995, Vintage, London

Pfeiffer, Ehrenfried; "Sensitive Crystallization Processes. A Demonstration of Formative Forces in the Blood", 1975, Anthroposophic Press, Spring Valley, NY, USA

Picoche, Jacqueline; "Dictionaire de l'Etymologier du Francais". 1973, Robert, Paris, France

Popper, Karl, R.; "Objective Knowledge. An Evolutionary Approach". 1972, Clarendon Press, Oxford

Prigogine Ilya and Stengers, Isabelle; "La Nouvelle Alliance", 1979, Gallimard, Paris

Rudnicki, Konrad, "The Cosmological Principles", 1995, Jagellonian University, Krakow, Poland

Searle, John, R.; "Is the Brain's Mind a Computer Program?", 1990, "Scientific American", vol 262, Nr. 1, p 20

Searle, John. R,; "Deux Biologistes et un Physicien en Quete de l'Ame", 1996, "La Recherche". Nr. 287, p 62

Shinbrot, Troy.; Grebogi, Celso; Ott, Edward; Yorke, James, A.; "Using Small Perturbations to Control Chaos", 1993, "Nature", vol 363, p 411

Siemens Jean-Louis, "Quelques Rèflexions sur les Cycles de l'Histoire Humaine", 1996, "Troisième Millènaire". Nr. 41, p 52

Steiner, Rudolf; "An Outline of Esoteric Science", 1996, Anthroposophic, New York

Steiner, Rudolf; "The Philosophy of Freedom", 1964, Rudolf Steiner Press, London

Steiner, Rudolf; "Towards Social Renewal", 1975, Rudolf Steiner Press, London

Steiner, Rudolf; "Anthroposophical Leading Thoughts", 1973, Rudolf Steiner Press, London

Steiner, Rudolf; "The Origins of Natural Science", 1985, Rudolf Steiner Press, London

Steiner Rudolf; "How to Know Higher Worlds", Anthroposophic Press, New York, 1994

Tetens, Johann Nicolaus; "Philosophishe Versuch", 1913, Verlag von Reuther & Reichard, Berlin

Thewlis, J.; Glass, R. C.; Hughes, D. J.; Meetham, A, R.; "Encyclopedic Dictionary of Physics", 1961, Pegamon Press, Oxford and London, New York, Paris

Thomas, Nick; "Rethinking Physics". 1996, "Newsletter Articles Supplement of the Science Group of the Anthroposophical Society in Great Britain". no 2, p 1

Thomas, Nick "Science between Space and Counterspace. Exploring the Significance of Negative Space", 1999, Temple Lodge Publishing, London

Uus, Undo; "Blindness of Modern Science"; 1994, Tartu Observatory, Estonia

Unger, Georg; "Forming Concepts in Physics", 1995, Parker Courtney Press, Chestnut Ridge, NY, USA

Young, Bob; "Science is Social Relations", 1977, "Radical Science Journal", no 5, p 65

Zurek, Wojciech H.; "Decoherence and the Transition from Quantum to Classical", 1991, "Physics Today", vol 44, Nr. 10, p 36

Zwicky, Fritz; "Morphological Astronomy". 1957, Springer, Berlin. Göttinen, Heidelberg, Germany

0-595-27960-0